A VENETIAN ISLAND

New Directions in Anthropology
General Editor: **Jacqueline Waldren**, *Institute of Social Anthropology, University of Oxford*

A VENETIAN ISLAND

Environment, History and Change in Burano

Lidia D. Sciama

Berghahn Books
New York • Oxford

First published in 2003 by **Berghahn Books**

www.berghahnbooks.com

First paperback edition published in 2006
Reprinted in 2006

© 2003 Lidia D. Sciama

The Library of Congress has previously catalogued a hardback edition as follows:
Sciama, Lidia D.
    A Venetian island : environment, history and change in Burano / Lidia D.
Sciama.
        p. cm. -- (New directions in anthropology ; v. 8)
    Includes index.
    ISBN 1-57181-920-7 (cloth : alk. paper)
    1. Burano (Italy)--History. 2. Burano (Italy)--Social life and customs. I. Title.
II. Series.

DG684.16.B87 S35 2002
945'.31--dc21   2002018402

British Library Cataloguing in Publication Data
A catalogue record for this book is available from
the British Library.

Printed in the United States on acid-free paper
ISBN 1-57181-920-7 hardback
ISBN 1-84545-156-2 paperback

CONTENTS

❧

LIST OF ILLUSTRATIONS

꧁❦꧂

Figures

Maps

Tables

Unless otherwise indicated, illustrations are by the author.

ACKNOWLEDGEMENTS

This book has evolved over years of research and reflection. I therefore have many people to thank, first and foremost the people of Burano, who welcomed me and were always ready to enlighten me on matters they regarded as implicit common knowledge. Their natural charm, sense of humour and generous hospitality made it a privilege to work with them. Various papers based on early fieldwork reports were presented at the Thursday seminars on the Social anthropology of Women – later Centre for Cross-cultural Research on Women – at Queen Elizabeth House, where discussion helped to deepen my then developing interest on gender aspects of Burano's society. The Identity and Ethnicity seminar convened by S. Ardener with Tamara Dragadze, Ian Fowler and Jonathan Webber, at the Oxford Institute of Social and Cultural Anthropology offered me valued opportunities to test my ideas in a context of stimulating and constructive criticism.

Many friends and colleagues encouraged me to publish the book. I particularly wish to thank Helen Callaway, Joanne Eicher, Dorothy Helly, Michael Herzfeld and Renée Hirschon for their sensitive appreciation of my work, their good-humoured criticisms and their invaluable suggestions. Also Peter Parkes, Robert Parkin, David Sutton and Jaqueline Waldren read early drafts of some of my chapters. Their insightful questions and advice greatly helped me to achieve greater clarity. Shirley Ardener's long-standing interest in and positive view of my work has been most heartening. I am grateful to the Gladys Krieble Delmas Foundation, whose grant at the very early stages of fieldwork encouraged me to persevere.

I greatly benefited by my institutional links with Somerville College, St Antony's College, Queen Elizabeth House, Wolfson College and, above all, the Oxford Institute of Social and Cultural Anthropology. Indeed I was much encouraged to publish this work by the recent publication of *Anthropologists in a Wider World* (eds. P. Dresch, W. James & D. Parkin, Oxford 2000), thanks to its writers re-examination of anthropological research and their openness to long-term, and sometimes unavoidably intermittent, fieldwork in an increasingly globalized world.

With an Italian fellow-anthropologist, Gianfranco Bonesso, then working in Burano, we exchanged views and ideas. Occasionally meeting mutual friends in the island provided both stimulation and pleasure, and it was of great interest to find that gender divisions in Burano also led to a degree of specialization in our ethnographic labours, given that my understanding of lacemaking and kinship attitudes paralleled his impressive knowledge of men's skills, and fish breeding techniques. Professors Glauco Sanga and Gianni Dore generously offered me much welcome academic hospitality at the University of Venice.

I also wish to thank Mr Fulvio Roiter, one of Venice's most accomplished and sensitive photographers, for allowing me to base the book's cover on one of his images of Burano.

My strongest gratitude goes to Dr John Campbell: his classic work on *Honour, Family and Patronage: A Study of Institutions and Moral Values in a Greek Mountain Community* (Oxford 1964) provided the main disciplinary basis for my research on Italian and Venetian society, and it is mainly due to his intellectual openness and generosity that this book was written.

I also extend my thanks to Dr Marion Berghahn, Dr Sean Kingston and Dr Jacqueline Waldren and to my anonymous peer-reviewers, whose readings undoubtedly helped to improve the Manuscript.

My husband and daughters read various drafts of my chapters and provided unwavering support, advice and encouragement.

FOREWORD

❦

This book is based on long-term fieldwork in Venice and in areas of the northern lagoon, mainly the island of Burano. My initial proposal, when I began my research for a D. Phil. thesis in 1980, was to conduct participant observation in Burano, while also maintaining a strong focus on Venice as the urban centre of which it is an integral part. In particular, I planned to analyse environmental problems in the lagoon and to examine ways in which inhabitants of Venice and Burano were affected by them and involved in their solution. Proposals put forward to repair ecological damage would illustrate relations between communities living in the island periphery and the city's politicians and administrators at the centre.

At that time my choice was undoubtedly influenced by changes and debates in anthropology, and especially critiques of structuralist studies. In particular, since the mid-1970s, critics of Mediterranean ethnographies have pointed out that attempts to describe societies as isolated and self-contained wholes had led to a proliferation of village studies in which little attention was paid to the dynamic relations of the areas observed with larger social contexts, whether cities, regions or nation states. The resulting overall picture was one of static social systems, hardly touched by change and modernity. Such concentration on village ethnographies, it was observed, was particularly surprising for a country like Italy, where urban values were pervasive and where migration from the countryside had been quite massive, especially since the Second World War (Boissevain 1975: 11; Crump 1975: 21–22; Davis 1977: 7–20; Gilmore 1980: 3; Macdonald 1993: 5–6, and see Just 2000: 20–28).

An initial theoretical problem was, therefore, the validity and usefulness of an urban/rural dichotomy – one as firmly rooted in anthropological tradition as it is in Italian culture. As we shall see, such dichotomous description did not fit my understanding of Burano, in some ways part of Venice, but in many other respects – and most importantly in its inhabitants' view – a village, separate and different from it. At any rate, as emerges from recent anthropological studies in cities, too clear-cut a distinction between urban and rural settings does not always fit ethnographic observations. For example, as Hirschon has shown in her path-breaking book on Kokkinia, an urban district between Athens and Piraeus harbour, in-depth examination of family, religious and in general moral and affective, attitudes, makes evident the continuity rather than opposition between rural and urban lifestyles and world views (1998: 232–5).

That observation is also fitting for Venice, where (leaving aside its foreign trading communities that in time became integral parts of the city) street names recall that significant groups of the working population came from its rural hinterland: workers in Murano's glass factories, as well as many of the city's bakers and domestic servants, came from Friuli, while smiths moved to Venice from Bergamasque villages (Trivellato 2000: 58–62; Sanga 1979). Herzfeld thus questions 'the wisdom of constituting a distinctive category and separate discipline "urban anthropology"' (Introduction to Hirschon 1998: xii).

Burano too proved resistant to a rigid categorisation based on an urban/rural (or non-urban) polarity: it is in many ways a village-like community, yet is part of the city's administrative and bureaucratic structures. Although it is undoubtedly very closely tied-in with the city, it could not be treated merely as a Venetian *quartiere*, because it emerged as an idiosyncratic and unique community, tightly bound to Venice and yet separate and keenly conscious of long-standing differences as well as dependencies and ties. Consequently, as I hope will become clear through my chapters, Burano eventually took centre stage, while notes on Venice are now kept in store, as part of future research on the city's historical centre. I should, nonetheless, emphasise that far from treating the island as a discrete and closed social entity, I focused mainly on its relations with the city, and their changes through time.

A challenging theoretical question concerning my position as fieldworker was readily prompted by my thesis title: *Relations between Centre and Periphery in the City of Venice. A Study of Venetian Life in a Lagoon Island.* Why, I was asked, had I defined Venice as the centre, Burano the periphery? Was I placing myself on one side of a power structure, whether cultural, political or economic, through which Burano was marginalised? The question was certainly thought-provoking, given that, like the urban/rural opposition, that of centre/periphery has been overlaid with polarities such as advanced/backward, modern/traditional, complex/simple and so forth, which clearly bring to mind ethical concerns and postcolonial critiques of anthropology. A conception of cities as the main centres of social and intellectual progress has deep roots in the discourses of Italy's dominant culture. For Venice, the point is best illustrated in early maps. For example, in the work of a Renaissance cartographer, Benedetto Bordone (1500), the city is at the centre of the lagoon, enclosed within a string of beaches, the Lidos; and, while in reality the beaches are straight, in the map they are drawn as an almost circular protective wall (map f. 1,2,3). By contrast, Burano is little more than a dot on the map's north-eastern corner; with its tiny bell-tower, it appears as a distant outpost of the city and the Church. Such stylisation considerably alters the lagoon's topographical reality (map f. 2), since, placing the city at the top of a hierarchy of human settlements, it clearly reflects an ideological stance of which present-day Buranelli are well-aware and sometimes resentful.

On reflection, while bearing in mind implications of this influential commonplace of Italian culture in terms of hierarchy and power, I have nonetheless found it unnecessary and potentially misleading to change my description of Venice and Burano as respectively centre and periphery – one I wholly share with the Buranelli's own characterisation of their island, based, as they say, both on topography and on culture. However, I have found it useful to make a distinction between the 'centre' as a space characterised by multiple elements of organisation (Hannerz 1980) and my 'frame of reference' as observer. Because I conducted research both in Burano and in Venice's historical centre – and often went from fieldwork to library – I clearly had a double vantage point, but in reconstructing the history of relations between inhabitants of the two areas and looking at their interactions and mutual stereotyping, my main frame of reference was Burano and my focus its inhabitants' viewpoints and sensitivities.

Far from describing Buranelli as the passive recipients of Venetian hegemony and economic power, I have emphasised their agency, as shown, for example, by their strong sense of independence, their capacity for hard work, the efforts of Burano's politicians to obtain better housing and medical services through democratic debate and, not least, by the women's virtual abandonment of unrewarding lacemaking and their rejection of strict family and social control.

To come back to my fieldwork, also other – more contingent – considerations led me to think (even against the wise advice of some of my teachers) that mine was a potentially rewarding choice. Personal reasons interacted with those of a disciplinary and academic nature, but were nonetheless integral to my work's development: I was born and educated in Venice, but spent most of my adult life in England. Indeed, my curiosity about that city's social life and history was undoubtedly sharpened thanks to many conversations with colleagues and friends in Oxford. When, for example, on being asked where I came from, I explained that I was Venetian, their response was frequently one of surprise. 'Why', I was told several times, 'I didn't think people actually *lived* in Venice'. This was usually followed by further questions, such as was Venice really sinking, was anything being done about it, and so forth, which made me feel as if I too had somehow been implicated in the culpable neglect of the city's environment. My fieldwork in Venice was thus also connected to my relations with persons in England, where I had lived since the 1960s, and this too raised theoretical questions that I can sum up as 'problems of representation'.

Partly, my interlocutors' interest was due to a rather unhappy and frightening circumstance – one experienced by Venetians as a 'total event' that had changed their attitudes and their discourses about the city: a violent flood in November 1966 (see below) had made evident both the city's fragility and the fact that its protection had been neglected by politicians and administrators and indeed gravely undermined through indiscriminate industrial development at the margins of the lagoon. Because of a coincidence in time with the period of political protest and environmental concern that culminated in 1968, the predicament of Venice was often cited as a foremost example of the evils of industrialisation and capitalist greed.

Throughout the debate that followed the floods – and in witnessing the sense of alarm, and the world-wide activities of dedicated committees in publicising the city's plight and gathering funds for its restoration – my response may have been one in which I identified mainly with Venetians: solicitude on the part of the international community was clearly due to the status of Venice as an object of beauty and a testimony to past political and artistic achievement. It was a concern with heritage, not always matched by true understanding of the history and the needs of the population. It was quite revealing in this respect that, with few exceptions, most historical or art-historical studies in English ended at 1797, when Venice's life as an independent city-state came to an end. Aspects of its day-to-day contemporary existence were then overshadowed by a heightened awareness of, and interest in, the past – or, in Herzfeld's words, the 'monumental past' (1991a). By contrast, to my knowledge, Venice had not yet been a subject of anthropological study.

Although I was well aware that no account can uncritically claim adherence to 'reality', I shared with many Venetians a feeling that a steady production of historical works, paralleled by ignorance of the population's circumstances, perpetuated a vision of Venice as a dying city, elaborated by Romantic and Decadent poets and writers, especially since its loss of political independence. Typical in this respect was Ruskin's description of the view from Torcello's bell-tower: 'There are no living creatures near the buildings, nor any vestige of village or city round about them.' Probably blinded by his long view of the sea, the 'misty land of mountains touched with snow', and Venice's 'multitude of towers, dark and scattered among square-set shapes of clustered palaces, a long and irregular line fretting the southern sky' (1867, II; 12) – and no doubt absorbed in the study of Torcello's Byzantine ruins – he then entirely failed to note the lively presence of nearby overpopulated Burano. The theme of representation, both of Venice by an international community of scholars, art-lovers and tourists, and of Burano, mainly by other Venetians, is a subtext I have not developed in this work, but one present in several parts of the book.

My position as 'native' observer also raised other theoretical problems: because I was born and brought up in Venice, my colleagues very soon asked, was I conducting 'anthropology at home'? (Jackson 1987). Or was mine an exercise in 'personal anthropology'? (Pocock 1977). But, while I fully appreciated that my choice was clearly a departure from a well-established tradition of long-term continuous fieldwork in some remote and isolated area, I did not experience the matter as a problem, either in practical or intellectual terms. After all, as Allen writes, pointing out 'the relativity of otherness', 'there is no cut-off point where otherness begins or ends' (2000: 245). Perhaps due to my Jewish, Venetian and Italian roots, I actually find it difficult to conceive of any society or community, albeit metaphorically, as 'home'. At that time a distinction between home/private and society/public seems to have been firmly in place in my unreflective use of language, and I still do not usually think of 'home' as anything other than a very intimate and personal place I associate with the family and the hearth, certainly not opposed to, but distinct from, community and country (Du Boulay 1974: 38; Hirschon and Gold 1982: 66;

Sciama 1981: 101). Such divergence in my Italian, as distinct from English, understanding of 'home' finds validation in dictionary glosses for, while in Chambers Twentieth Century Dictionary (1949) we find home = 'one's house or *country*' (my italic), Zingarelli's Italian dictionary (1962) gives simply 'walled edifice built for dwelling, family, or persons who dwell together as family'. In other words, the meaning of the Italian *casa* does not extend to country, village or nation.

However, my unease about the very idea of 'anthropology at home' need not be explained only in terms of lexical difference; as I hinted above, it may be due partly to personal history: I left Italy in my early twenties, settled in England on marrying in 1960 and studied anthropology in Oxford as my second education from 1970. Following Strathern's redefinition of 'anthropology at home' as 'auto-anthropology', that is, 'anthropology carried out in the social context which produced it', my field could not easily be described as 'home' (1987: 16–17). By the time I started my research, my long residence in England had changed my perspective and, despite frequent visits, had created a distancing from Venice and Italy, while other factors that contributed to my difficulty in defining my research as 'at home' were undoubtedly the limitations of my early social experience, due to boundaries of education, place of residence and schooling, as well as the fact of having grown up in the 1940s as a member of a minority in a then divided and oppressive society.

All the same, I certainly would not wish to deny or renounce a strong sense of belonging that still ties me to Venice: while at the beginning of fieldwork my knowledge of Venetians, especially those spread through the lagoon's islands, was limited, I knew my way around the city's libraries and archives, and I occasionally benefited from the presence of some old acquaintance in relevant administrative or political offices. Above all, having grown up as a Venetian speaker, I was easily able to cope with my informants' diglossia, or their outright reluctance to speak in Italian. In sum, my position as fieldworker is probably best described in Peter Loizos's observation about the slippery nature of categories like 'outsiders' and 'insiders'. 'Insiders who become anthropologists in their own societies [he writes] have obviously undergone a trans-formation. [They have] come out of a single culture and become bi-culturals." (1981: 170–71).

As will appear from references to my fieldwork in various part of this work, I did not develop a strictly worked-out methodology; indeed I found serendipity was a better guide than a pre-established set of rules and guidelines. For example, at the outset I had no interest in lace, I did not know that lacemaking, with all its implications in terms of gender, honour and sexuality was quite as prominent a factor in Buranelli's sense of their history and self-image as I eventually learned from my informants' accounts. Because I hoped to be accepted in the island in complete independence from any political party or organisation, I did not make contact with agencies or persons who might have helped me to settle there, but I am still not certain whether this had the desired effect, given that living with a family itself made it difficult to get close to persons of different political orientations and views.

As well as observing Buranelli's everyday lives and trying to identify their interests, analyse their attitudes to family, history, religion and authority, in light of their changed living conditions and outlooks, my attention was inevitably focused on the theoretical questions that had dominated the anthropology of Italy and of the European Mediterranean since the 1950s, in particular the 'honour and shame' complex. A rethinking of 'honour and shame' through the analytical lens of gender led to a view of both terms of the dichotomy as more nuanced and less gender-biased than I might have assumed on the basis of earlier ethnographies in comparable areas.

Finally, another question that runs through the work is that of the Buranelli's construction of their collective identity, mainly based on their strong sense of place and their view of their society as a community structured and held together by its intricate and numerous kinship links and its people's use of kinship idioms.

The Writing

Writing about relations between Venice and Burano poses problems of style, and, in particular, of priorities and relations between different parts. Given the complexity and inter-relatedness of topics such as honour and shame, religion, economics, politics and modernisation, themes could not be tidily confined to different chapters. For example, my section on religious practice and belief is relatively short – not because religion is not central to Buranelli's lives, but, on the contrary, because it so pervades all aspects of Burano's society that its discussion is inevitably present in my chapters on history, kinship, honour and lacemaking.

Although I was quite mindful of the critiques of the writing styles, subject divisions and aspirations to objectivity of earlier ethnographies – and well aware that my chosen field would have offered ample opportunities for experimental postmodern writing – I have mainly followed a conventional style. Because of the changing, elusive, and sometimes difficult to grasp social reality, I found that a formal treatment, especially in my chapters on kinship and 'honour and shame', helped to bring to light coherent patterns and structured regularities.

Given that environmental problems and dilemmas which characterise Venetian life – in particular, the contrast between conservation and change – are matters of concern and personal involvement for most people both in Burano and in the historical centre, I have introduced my first chapter with a brief description of the lagoon, which is their common geographical setting. In the second part of the chapter I have described the island's housing problems – for its inhabitants the most pressing of all environmental issues and the reason why numerous Buranelli, especially young couples, move to the Venetian hinterland. Buranelli's discussion of housing illustrated aspects of their lifestyles and residence patterns, while their expressions of anxiety when faced with the choice between seeking modern accommodation elsewhere and remaining on the island, despite its remoteness and its poor housing conditions, revealed their strong sense of belonging. As a dramatic example of their past poverty, poor housing was also the most obvious material evidence of a lack of assistance from the city and the state, and was a central issue in Buranelli's political debates in the 1970s and 1980s.

In Chapter Two, 'A sense of history' I report some of my informants' perceptions of the past, based both on known accounts, whether recorded or orally transmitted, and on their awareness of gaps and lacunae in what they would otherwise envisage as a continuous linear narrative. In that context, 'history' includes 'the past' and its 'written or orally transmitted records', as well as 'memories', while 'a sense' connotes knowledge and experience, but also fantasy, by which Buranelli construct their own metahistories. Narrative segments are therefore introduced in my text as flashbacks mostly prompted by questions developed in cooperation with informants in the field.

Direct quotation of descriptive and historical writings on Burano have been introduced, not only for their information value, but also to convey, in the writers' own voices, representations and views which in turn had a bearing on the islanders' self-images. One of my informants, for example, maintained that the portrayal of Burano's fishermen as comic villains in Renaissance comedies had considerable influence in developing their self-mocking and ironical vein. As several of my informants stressed, and as appears from literary evidence, throughout history Venetians and Buranelli have mutually emphasised their differences, while, at the same time (paradoxically) recognising their common roots. Indeed, writing about relations between Buranelli and inhabitants of Venice's centre, or in general those Buranelli regard as their 'others', naturally involves the issue of stereotyping and image-making. Descriptive accounts of the island at different periods, therefore, illustrate the writers' attitudes to rural, remote and peripheral areas – which were often equated with 'the primitive' in the eighteenth century and with backwardness and underdevelopment in the nineteenth to the mid-twentieth centuries. They thus cast some light on aspects of Italian social thinking.

One of the main themes in Chapter Two, but one present also in other parts of the work, is that of the embarrassment of historical riches, and, conversely, the sense of deprivation where documentary evidence is poor or entirely lacking. At the time of my fieldwork a superabundance of historical and literary writings on Venice was experienced by inhabitants of the historical centre as a weight, and there was a growing awareness that, when coupled with strict conservatism and added to the severe constraints posed by the city's environment, history could have a paralysing effect. By contrast, Buranelli, puzzled by obscurities in accounts of their community's beginnings, often deprecated the fact that their island had long been ignored by historians. As a result, speculation about their early origins was open-ended and indeterminate.

In my chapter on kinship, I have described patterns of residence and household organisation and have examined uses of kinship terms, as well as traditional jural rules and notions about reproduction. Older peoples' narratives of their early years illustrate modes of socialisation that clearly contrast with recent practices and show significant changes in family and gender relations, in particular in the exercise of authority. I have described uses and attitudes to nicknames, and have looked at ways in which kinship connections and networks link people resident in different islands and ways in which kinship shades into friendship at the outer edges of the kindred.

A long time-dimension was required in my chapter on honour and shame, because of the long history of both notions and their prominence as moral and behavioural complexes. Much has been written on the subject throughout the 1980s and 1990s, and some of my general criticisms of earlier works have been admirably covered, especially by Herzfeld, Llobera and de Pina Cabral, while various aspects have been discussed by Du Boulay, Hirschon, C. Stewart, Gilmore, Brandes, A. Lever, F. Stewart, Just, Goddard and others. However, I have concentrated on 'shame' as an experience and a concern of women – one that in earlier works was not treated as extensively as was male honour. As a sensitivity to censure and a fear of transgression instilled in early childhood, shame has undoubtedly been a strong factor in limiting the range of occupations open to women and maintaining rigid gender boundaries. Lacemaking in Burano, the topic of Chapter Six, is a particularly strong example.

The history of that craft shows how secular ideas about honour are supported by Catholic attitudes to gender and to women's sexual morality, while in the past a fear of shame was traditionally instilled at a very early age and went hand in hand with the teaching of the first lace stitches. Discourses about lacemaking, the main professional occupation of Burano's women, in which the craft and its history were overlaid with religious and symbolic meanings, as well as the Church's influence over its organisation, strongly reinforced patriarchal attitudes. By contrast, recent developments, especially since the 1960s and the closure of the old Lace School, are examples of significant change in Burano's economic and political life, as well as in gender relations.

In the last chapter I have returned to the ethnographic present (by now recent past) and have described particular moments in Burano's relations with Venice. Although the island was excluded from industrial development in the lagoon, and was thereby increasingly marginalised, its inhabitants too are now suffering the negative consequences of environmental damage. Their interactions with the Venetian municipality, therefore, are often tense and sometimes give rise to lively confrontations with the city's representatives. Debate is, as ever, conducted with a high consciousness of Burano's peripheral nature, and a keen awareness of long-standing connections, as well as resentments and memories of neglect and unanswered claims.

Most of the themes in this book – Buranelli's attitudes to their past, sentiments and behaviours in kinship relations, changing views about honour and women's work – are also relevant to the main issue of relations between Venice and Burano. Different threads come together in the last chapter, as in their political discourses and their negotiations with the Venice commune, Buranelli express their distinctive collective identity.

A Venetian Island

Map 0.1 Benedetto Bordone, 1528. Venice and the Lagoon. Oxford, Bodleian Library.

Map 0.1 Detail, Burano, Torcello and Mazzorbo.

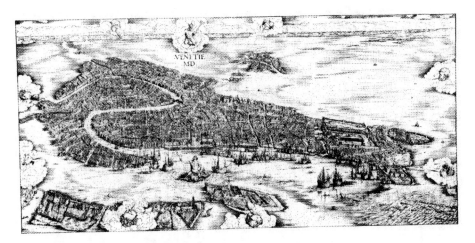

Map 0.2 De Barbari 1500. Original in Venice, Museo Correr.

Map 0.2 Detail, The islands of the northern lagoon.

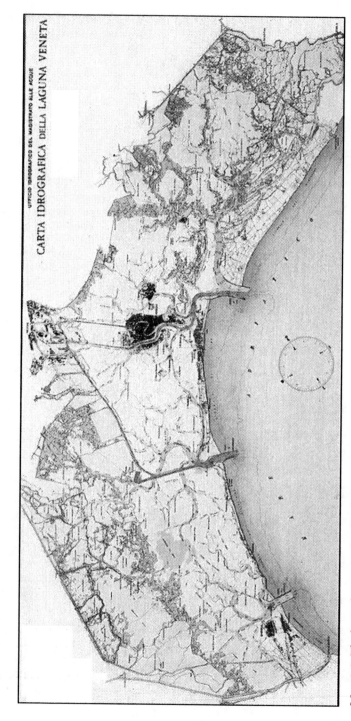

Map 0.3 The lagoon's topographical reality.

1
BURANO, VENICE AND THE LAGOON

From the mouths of the Adige to those of the Piave there stretches, at a variable distance of from three or five miles from the actual shore, a bank of sand, divided into long islands by narrow channels of sea. The space between this bank and the true shore consists of the sedimentary deposits from these and other rivers, a great plain of calcareous mud, covered, in the neighbourhood of Venice, by the sea at high water, to the depth in most places of a foot or a foot and a half, and nearly everywhere exposed at low tide, but divided by an intricate network of narrow and winding channels, from which the sea never retires. In some places, according to the run of the currents, the land has risen into marshy islets, consolidated, some by art, and some by time, into ground firm enough to be built upon, or fruitful enough to be cultivated: in others, on the contrary, it has not reached the sea level; so that, at the average low water, shallow lakelets glitter among its irregularly exposed fields of seaweed. In the midst of the largest of these …the city of Venice itself is built, on a crowded cluster of islands.[1]

John Ruskin, *The Stones of Venice*

The island of Burano, about six miles from Venice, is half-way between the marshy northern coast and the tip of the peninsula of Cavallino, which forms the lagoon's outer boundary. It is part of a small estuary, with Torcello on the north and Mazzorbo, to which it is joined by a wooden bridge, on its western side. About two miles south is the monastery island of San Francesco del Deserto, and just beyond, stretching westwards, is the large agricultural land of Sant' Erasmo (map 1.1).

What mainly distinguishes Burano from other parts of the Venice commune is the fact that its area, over fifty square kilometers, is larger than those of all other districts, except Cavallino. Most of it, however, is an expanse of water, while its dry land is barely more than half of one square kilometer, only 1.1 percent of the total[2] – a fact which explains both the sense of Burano's remoteness, and its

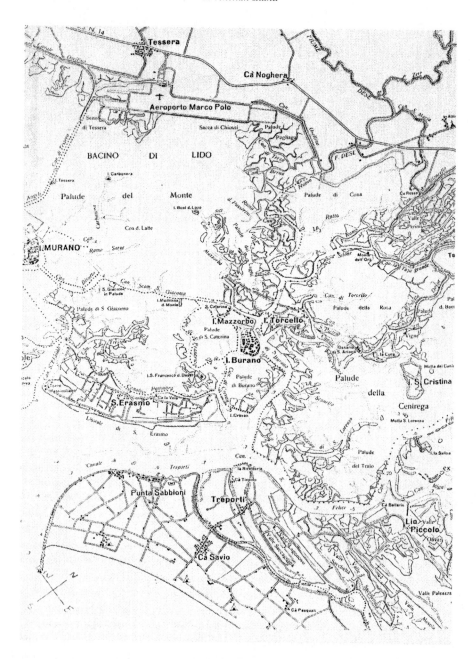

Map 1.1 Burano's waterways from the Dese river to the Canale di Burano.

inhabitants' statement that ' in the lagoon are [their] fields' – one better justified when fish was more plentiful and less endangered by pollution than at the present time. Indeed, no account of Burano's (and Venice's) economy, history or life-style is complete without some understanding of the lagoon. Therefore, before I proceed any further with my description of that island, a brief account of the nature and configuration of the lagoon will help to explain some of the problems and the constraints of the Venetian environment.

Viewed from the air, the overall shape of the Gulf, which spreads alongside the western edge of the Adriatic sea, can be compared to a crescent moon, the shorter side being the beaches, while the more curved and longer side, rendered complex and rugged by numerous river mouths, is the shore of the terra firma. Its area is about 500 square km, and its boundaries are the rivers Sile and Brenta, respectively on the north and south. On the west a series of canals mark the line between the lagoon and the hinterland, while its eastern borders are the littorals of Jesolo, Cavallino, Lido, Pellestrina and Sottomarina, which separate it and, to some extent, protect it from the waves and storms of the open sea.

In origin the lagoon was a delta region, but, as Ruskin's passage explains, its present state is the result of long-standing interplay between the sea and the numerous rivers that issue in the Venetian gulf: it gradually evolved as solid detritus carried by rivers, on meeting the contrary tidal waves of the sea, was redistributed and settled in places where the soil of the sea is gently sloping and the waters are shallow, thus forming the beaches which now enclose its water mirrors. The string of beaches, however, remained naturally open in three places (*bocche a mare*) which allow the periodical ebb and flow of sea water. Rather surprisingly, the lagoon actually consists of three distinct basins whose waters and tidal movements never mix, although they are not separated by any barrier.

The Venice lagoon, like other comparable environments, is constantly in a state of transition, since its evolution is tied to numerous and sometimes contradictory factors such as the rising of the sea level, which is especially due to phenomena of glaciation, the vertical movements of the soil and changes in the patterns of settlement of the fine sediments carried by the rivers, which may in turn cause changes in the depth of the lagoon basin. Its conservation therefore depends on the maintenance of a dynamic equilibrium between all these factors, since it may otherwise be destined to become a tract of sea if the marine erosion prevails, or to become dry land, if the detritus carried by the rivers becomes prevalent.

Thus, far from being a natural environment, the lagoon has been very strongly shaped through human intervention. Its present configuration is largely due to complex engineering works, mainly devised to keep out the river waters that debauched into the sea through Venice or its vicinity, and to ensure that alluvial deposits should be spread as far as possible. To that end, the waters of all the rivers that flowed into the lagoon were deviated through artificial canals, or channeled into river beds on the north and the south, so that they could flow directly into the Adriatic without touching the city.

Due to those operations, gradually carried out over the last seven centuries, the environmental regime of the lagoon was changed: the whole area of the terminal portion of the deviated rivers was filled with sea water and the boundary of salinity shifted towards the terra firma and the internal coast. Consequently, the network of ancient river beds was much changed. In the vicinity of ancient river mouths, where the action of the tides was strong and dynamic, the depth of the lagoon was maintained, or even increased, but in areas further away from the rivers and the sea it was progressively rendered shallow, while on the flat coastal plains there formed shallow pools that could be invaded by sea water only at high tide.

As Bellavitis points out (1985: 4), this phenomenon, already noticeable in maps of the early fifteenth century, became increasingly evident. Indeed, the lagoon of today, viewed from above or from hydrographic maps, looks like the efflorescence of a large tree with very sinuous arms. Its trunk, rooted in the sea, is the channel that leads to the port, and which, touching the beaches of Lido and Cavallino, allows the sea waters to enter the central and northern part of the lagoon where Venice, as well as the islands of Burano, Mazzorbo, Sant' Erasmo and Torcello are situated. A similar shape also characterises the two southern channels of Malamocco and Chioggia. Due to the hydraulics works I have briefly described, the basin's original 'river-delta' and the 'sea-lagoon' matrices are no longer present in equal proportions: once the river waters were channeled into their new beds, the delta matrix was reduced to a morphological substratum of the sea and the lagoon, which now dominate the environment. However, as Bellavitis writes, the imprint of a double matrix, a deltaic and a maritime one, can clearly be discerned if we observe the lagoon's settled areas in the two extreme conditions of its tidal rhythm. During high tides, and especially against the light, they appear as compact bodies emerging from a vast expanse of water, but at low ebb, they look like inlays in a vast swamp, built on the dry portions of a meandering river delta. The existence of this double image and double structure has given rise to a literature and iconography of Venice which are similarly double, while some of the difficulties met by administrators in the solution of its environmental problems also find their source in this double structure.

Venice too, like the lagoon, is described by geographers as the result of interaction between nature and artifice – one in which, perhaps to a greater extent than in other cities, human ingenuity has overcome the problems posed by an unusual and difficult terrain. Like other so-called 'delta-cities', Amsterdam, Calcutta, Saigon and St. Petersburg, it was built largely on artificial ground, but it differs from those cities in that, as we have seen, its lagoon ultimately prevailed over an original deltaic morphology (ibid.).

As one flies towards its airport on a clear day, Venice looks like a compact and elongated island-city, comfortably nestled in its gulf and joined to the mainland by a bridge and a railway over just a few miles of lagoon water. Ruskin's incisive description of Venice's ecological datum, 'a crowded cluster of islands', clearly evokes a place which is at once diversified, intimate and sheltered. The simple fact of being built on one hundred and twenty islets, joined by four hundred bridges over canals of various depths and widths, and the constraints this inevitably poses, has naturally favoured the maintenance of an idiosyncratic local character within each one of its neighbourhoods, so that even a casual visitor obscurely perceives how each area still retains its own architectural and social identity, related to its history, its parish, school, and other institutions. The diverse character of Venetian neighbourhoods, however, becomes confused and almost imperceptible in the centre of town, where an active commercial life and the coming and going of people to and from school, office, or church make it almost impossible to be clearly aware, except through a keen training of one's historical and sociological observation, that one is constantly moving from one city quarter, and one parish or neighbourhood, into another.

As we move away from the central area, between Rialto and San Marco, different parts of the city seem to take on more marked characteristics, while, if we go even further and take a boat to the outlying islands, we find that each one has its own well-defined character. Indeed, the geographical formation of the city lends itself to a territorial division of labour, responding to the most articulate and well-developed classificatory principles on the part of its people, and, in particular, of its past rulers and planners. The best example is possibly that of Venice's psychiatric and isolation hospitals. For, while until recently most cities have isolated their mentally ill, as well as those affected by contagious diseases, in some remote and little used area, in Venice, where a number of small islands just a few miles away from the centre, readily offered themselves for such uses, the separation was naturally to become all the more drastic and poignant. A spatial division and distribution of social functions may at the same time convey a pleasing sense of visible order, making sharper distinctions between the healthy and the ill, the sane and insane, the male and the female, or the true citizen and the temporary resident or trader. It may be no coincidence that Venice was the first European city to have a Jewish Ghetto, closed by gates in 1516, and to settle traders of different nationalities, especially Greeks, Dalmatians, Armenians, Turks and Germans, in their own quarters.

The variety of Venice and of its surrounding islands may be one of the features that make the city so appealing and memorable to its visitors, hurrying through a day trip and learning to associate each island with a particular trade, function, historical event, or religious order. While the centre of the city has achieved a unity and functional interdependence which make it difficult to disentangle quickly its separate neighbourhood or parish histories, the outlying islands still retain their somewhat different characters, and, to some extent, their idiosyncratic forms of social life.

As we shall see (Appendix 1), the Venice commune includes large inland areas, but, given the diversity of its various territories, it has been subdivided into three main administrative units: the historical centre, with four city quarters, the estuary, with five quarters, and the terra firma, with nine. These offer the main bases for comparison, and indeed they show the most striking differences in terms of environmental structure, as well as census, development and migration figures. Although all areas are interdependent, and indeed remarkably cohesive in economic as well as cultural terms, their histories have been different and in many ways contrasting, owing to their different ecological characters. In particular, over the last fifty years the historical centre and some estuary islands, including Burano, have suffered the most thorough neglect and obsolescence: decay of their urban fabric, environmental pollution, and an increasing commercialisation of tourist activities have been paralleled by a severe loss of population.

Burano

In contrast to the narrow strips of Sant'Erasmo and the Lidos, where the soil is mainly formed of straight and extended sand dunes deposited by the sea, Burano's overall shape, moulded through time by the currents and filled with the silting of rivers, is roughly rounded. As appears from maps and aerial photographs of the northern lagoon, a string of lowlands, criss-crossed by meandering rivulets and canals, which are usually in evidence at low tide, seems almost to form a bridge from the internal coast to Burano and all the way to Tre Porti and the Cavallino peninsula. Although the island is very close to swamps on its southern side, it is by no means marshy. On the contrary, its north bank sides on a deep channel, the Canale di Burano, a lively branch of the river Sile, and part of a network of waterways that connect areas of the internal coast, such as Tessera, now Venice's airport, with the Lido and the sea (map 1.1). It is a densely settled island, with a population of 5,208 at end of 1981 (now considerably lower due to people moving out and to a diminished birth rate).

Buranelli's historical memory and their deep-rooted sense of their own marginality were shaped during centuries of Venetian rule and, no doubt, reinforced by a long-standing vision of the superior power, civilization and wealth of cities, and in particular Venice. Indeed, while they proudly acknowledge their Venetian identity, they often say that an even stronger attachment and sense of belonging ties them to their island, to which they variously refer as their village or neighbourhood. The history of Burano's administrative relations with Venice has been a changing and sometimes a complex one. In Roman times it was administered as a *vicus*, or outlying village, of Torcello; under the aegis of the Venetian Republic, it was classified as a hamlet (*contrada*) and administered from Torcello's town hall (*podesteria*), but it eventually became the main administrative centre of the northern lagoon, when Torcello's population was decimated by malaria and almost totally abandoned.

In the late eighteenth century, after Napoleon's invasion and the loss of Venetian independence, Burano became a commune and part of a District. It continued to have a somewhat separate communal life under the Austrian

(1915–1966), and then Italian governments, but it was finally joined to the city's Municipality in 1924, and is now classified as one its *quartieri*. Despite political and administrative changes, however, Buranelli have always looked to Venice as the main urban centre, of which their island is in all ways a peripheral village. From their point of view, Venice is a true metropolis, with hospitals and out-patient services as well as high school education, and it is often in Venice, or more recently Mestre, that they like to do their more important shopping, even though it takes three quarters of an hour to reach the city by motorboat. Above all, however, Venice was traditionally the economic centre where the islanders sold their fish and lace, found employment as sailors, boatmen or shipbuilders in the city's arsenal, and where some of the women made a living by collecting the debris of Venetian luxury to resell second-hand. Indeed the cry of Burano's women seeking to buy 'rags and bones', O*ssi, strasse, feri veci. Roba vecia da ciapar bessi!* was one of the familiar sounds of my childhood.

The public boat from Venice to Burano leaves roughly once every hour from the city's sombre north-eastern side, the Fondamente Nuove. As one enters the boat, one can immediately perceive a different atmosphere from that of water-buses in the historical centre; tourists are relatively few, since a special tourist boat departs from the San Marco quay, so that large groups of foreign visitors, following their own routes and timetables, remain peculiarly separated from the locals. Consequently, the boat to Burano is mostly used by commuters, schoolchildren and groups of glass workers going to or from Murano.[3]

On boarding the boat, therefore, one at once had the feeling of having entered a social world more intimate and relaxed than that of the city. The journey, about forty-five minutes, is sufficiently long to allow people to turn to some diverting activity after a long day's work or shopping. People often arrive in groups of two or three, and as they choose their seats on the boat new groupings are formed. School boys and girls tend to sit together, later to separate, when the boys start a game of cards, while the girls take out their knitting or crochet work; two or three young couples may carry on a general conversation, while middle-aged women and men tend to form smaller groups (figures 1.3 and 1.4). Young mothers usually travel with a relative or friend and, should they be alone with a small child, it is rather unlikely that a passenger does not sit next to them and exchange a few remarks. The boat first stops at the cemetery island of San Michele and at Murano, then follows the longest part of the journey over the stretch of lagoon from Murano to Mazzorbo. After touching Mazzorbo, the boat goes across to Torcello, and only stops at Burano on its way back. Those who are impatient to reach home, therefore, leave the boat at Mazzorbo to walk hastily along its bank and over the long wooden bridge which connects the two islands.

As one of Venice's historians observed, 'Burano is like a small-scale version of Venice, with its internal canals, its embankments and bridges' (Miozzi 1968, III: 202). Like Venice, too, Burano is built on a number of islands: San Mauro, San Martino Destro, San Martino Sinistro, Giudecca and Terranova, and each of these, according to its inhabitants, still maintains its own different character and some traditional forms of speech (map 1.2). At a first view from the boat's landing, the island presents a large open space, somewhat like an ordinary quay, where periods of silence and moments of lively noise and hurried traffic alternate according to the boat's timetable. Four large stones on the right, as one alights from the boat, create a notional boundary between the busy passageway and a wide, grassy common, shaded by trees, where boys play their noisy football games, visitors have their picnics and young couples linger about at dusk, on their way to a walk in Mazzorbo.

In contrast to the right side with its open common, the left side of the quay is built up with small, colourful houses, most of the ground floors of which have recently been turned into lace and souvenir shops. A neat row of brick and red-painted houses and a small café, its tables hemmed in between two shops, altogether give an impression that we are in a well-developed, however minute, urban space. The street narrows for a short stretch, then opens again to reveal a more complex and varied view of quays and terraced houses, and one is now faced with a wide canal crowded with small coloured boats, moored on both sides. The left-hand quay leads to Burano's main centre, the Via Baldassare Galuppi, a wide, lively street, which, until it was filled in 1885, was the island's main waterway, equivalent in terms of its function, if not of its size and grandeur, to Venice's Grand Canal (figure 1.1). The Via Galuppi, which is now comparable to any small town or village high street, has a number of restaurants, cafés and tourist shops, while the meeting rooms of the Socialist, Christian Democrat and Communist parties, well placed at its centre, are only separated by two or three houses. The main thoroughfare is connected to its surrounding areas by narrow streets, which create an overall impression of a constant alternation of light and dark and of wide open spaces and narrow passages, or intimate, and almost private, internal courtyards.

The Via Baldassare Galuppi opens on to Piazza Galuppi, a very large uncluttered space, graced by a well at its centre. There stands Burano's main church, San Vito, 'a large austere building mainly of the sixteenth century' with, at its side, the small oratory of Santa Barbara, where a painting of the crucifixion is thought to be an early work of Tiepolo (Honour 1965: 235). Viewed from the outside, and compared with the minute size of its neighbouring houses, the church appears almost disproportionately large. Leaving the church and the piazza behind, one comes to Burano's south-eastern shore, facing the islands of Sant'Erasmo and San Francesco del Deserto, the most peaceful of views, which evokes a lagoon only settled by fishermen, agriculturalists and pious establishments. Because San Francesco is not connected by a public boat service, at the time of my fieldwork a very old boatman would ferry people across for a small fee.

By contrast, large groups of tourists are often disembarked at that end of the island and are guided through its square and main street following a direction opposite to that of the people who arrive on the public boat. The route I have described is therefore known to most tourists; it has a natural rhythm and direction of its own and it seems to leave the stranger with an impression that he has truly stepped into a different social world from that of Venice or other islands. Everything except the church looks smaller, and the fact of having completely crossed the island in about ten minutes gives rise to a feeling of quaintness and intimacy that may even lead to an exaggerated notion of its smallness. As one wanders off the main route, one clearly sees that, especially on warm days, life is led outdoors, and domestic tasks such as washing, grilling fish, sewing and lacemaking, are mostly carried out in the streets and the *corti* (figures 1.5 to 1.16). But, while this may give outsiders a feeling of ease and openness, equally it may build up to an apprehension that they are intruding into a space in which they have no real place or function.

Buranelli always welcome their visitors and are noted for their courteous and urbane manners. They are always willing to let themselves be observed and photographed, but they have taken no initiative in offering any organised hospitality for more than a few hours. The island's restaurants are large and efficiently run, and are also known to offer the best in the way of seafood and pastas, but very little overnight accommodation can be found beyond that of two or three bedrooms above a tiny old-fashioned inn. The largest restaurant, Romano, has a very important place in the social history of Venetian art at the turn of the century, when several distinguished landscape painters used to gather in the island.[4] They would generally have meals in Romano's spacious dining room and drink with the local men. A few privileged persons were actually accepted to stay there, mainly thanks to their well-established relations of friendship and mutual trust with the host.

Romano's restaurant is also one of the few places where local and non-local life meet —without really mixing – given that, as well as offering the best of Burano's cuisine, it specialises in wedding feasts, convivial parties of rowing champions and election meetings. In general, however, according to several of my informants, Buranelli prefer to keep their island to themselves, so that the limits of hospitality are well-defined and the most fitting conclusion to a brief involvement with outside visitors is the latter's return journey on the boat back to Venice.

The Fieldwork

I began my fieldwork in Burano in 1982 with a project on housing. I then proposed to inquire into the use of house and village space, the different attitudes of family members to the home, and people's needs and desires, particularly in view of government promises to allocate funds for new economic housing. Protests over housing conditions have been very frequent throughout the country, but they gathered strong momentum after 1968. In the large industrial cities of the north, especially Milan and Turin, large numbers of southern immigrants who had moved there since the 1950s and 60s were having to pay very high rents for inadequate accommodation, while throughout the countryside peasants still dwelled in decaying farmhouses deprived of any modern comforts.

In Venice's historical centre, where, as in other ancient cities, many houses were naturally time-worn, the protest was compounded with concern about pollution and the dilemmas of industrial development in Marghera. A rapid growth of tourism had led to a noticeable increase in the commercial use of accommodation and escalating prices, which made it very difficult for new families to find adequate homes. As a result, since the 1960s and 1970s the whole of insular Venice has seen a steady exodus of its population. The housing problems that caused people to move away from Burano were in many ways the same as those of the historical centre; its dwellings, however, summed up both the deficiencies of urban environments and those of ruinous housing in rural areas, where building had always been very poor and long neglected. To drive people away from the island was not only the high cost of restoration, but also the delays and constraints caused by stringent conservation rules.

Most of the people who left insular Venice moved to Mestre, Marghera or other nearby inland areas. In the same period, however, media interest in the discomforts and alienation of life in large noisy cities, where motor traffic seemed to prevail over more human dimensions of living, had led to a widespread feeling that living in Venice and its nearby islands was, after all, quite desirable. Uncertainty, and sometimes conflict among family members about choice of residence, had given rise to an ongoing debate about the advantages of Venice's, and Burano's, lifestyles and amenities, as against the hardships of life in the terra firma. My proposal, therefore, was to assess what people valued most: modern high-rise housing in the Venetian hinterland or the sense of freedom and open-air living of the island, the opportunity for boating and fishing, as well as the reassuring feeling of community, which many Buranelli claimed to greatly cherish.

By coincidence, a few weeks after my arrival a number of architects came to Burano to conduct a general survey of its housing. As several people assumed that I was a member of their team, I inevitably had to share in the mild hostility and suspicion that often greet Venetian officials. However, their presence also contributed to making me less conspicuous. At first I too was affected by the very problem I was proposing to study: my attempts to find a home completely failed and I was forced to commute from Venice. This was very much against my intentions, (it certainly was against the fieldwork imperative I had fully absorbed during my training) but it appeared quite natural and acceptable to the Buranelli – in so far as my being there was natural and acceptable at all. As they explained, except for a few autumnal landscape painters, who traditionally spend a few days or weeks in the island, they were not really used to people staying overnight. They generally expected outside personnel, such as their schoolteachers, social workers, and even their neighbourhood policeman, to commute daily to work – a fact which some of them resented and at the same time took for granted. Some people explained in self-deprecating terms that nobody, who had a home elsewhere, could possibly *want* to stay in Burano, even temporarily. Others said that, with their shortage of housing, they certainly would not wish people to settle there. At any rate, from my point of view, commuting was itself a form of adaptation to local ways.

During the initial stages of fieldwork I therefore concentrated on observation of the island's buildings and streets, and generally tried to identify themes of particular relevance to Burano's population. Eventually, a home was found for me by a Venetian woman lawyer, who persuaded one of her clients to offer me hospitality. Their relationship was itself an interesting example of a many-stranded and intricate web of common interests, obligations and counter-obligations, which, as I later found, was not unusual. I was thus offered accommodation in the family's large, musty, and little-used parlour, and, while this clearly implied a lack of privacy and offered little opportunity to write, it plunged me into a situation in which a great deal of learning about my hosts and their neighbours and friends was almost inevitable.

While my initial difficulties in justifying my presence on the island had been somewhat lessened by the arrival of architectural surveyors from the city with whom I was sometimes associated, after they left my continued presence did arouse some curiosity. As I told them, I was there 'to work', yet, without any known aim or obvious occupation, I sometimes appeared to be quite at a loss for something specific to do. In order to explain the nature of my work, and give a version of it that would actually register assent, I said that, as well as inquiring into housing conditions, I intended to record the dialect, with some of its typical expressions and narratives, and to learn about local customs and history.

I conducted the fieldwork mostly in dialect, but my unmistakably Venetian accent was sufficiently different from Burano's speech to give rise to observations on the varied nuances of that idiom, and on the social significance of language in general. Several people assumed that my habitual speech must have been Italian, therefore, given their keen awareness of the social and hierarchical implications of language, I soon learnt to tune in to the choices and preferences of my various interlocutors, and I readily spoke Italian with those people who evidently thought that Italian was more appropriate, effective, or decorous. That it was preferable to do so was particularly brought home to me when a young man, after taking his little boy to be examined by a doctor in Venice, told me how much he had resented the fact that the doctor had spoken to him in broad dialect, as if he thought him incapable of communicating in Italian.

As soon as suspicions that I too, like some of the teachers, social workers, or nuns working in Burano, might have been there to patronise, teach, or catechise the islanders were entirely dispelled (and my position of dependence as a researcher and a guest certainly contributed), Venetian became the normal speech. In my experience, conversations in that dialect were usually the most enjoyable, inasmuch as they allowed me fully to share in the islanders' warmth and good humour, while the fine shades of difference between Venetian and Buranello provided sufficient grounds for socio-linguistic observation. Indeed, mention of the dialect often touched on the islanders' feelings of being different, while questions about their history seemed to reveal some curious blanks, which they always acknowledged and somewhat resented, and to arouse their curiosity and desire to know more. For the family with whom I lived, my being 'English' and Jewish, as well as Venetian, seemed to confer on me some added dimensions, so that I too was an object of curiosity: differences thus opened the way for comparisons and for a search of commonalties that made the fieldwork all the more rewarding – or, at the very least, relieved potential guilt feelings at having to inquire into their opinions and beliefs, and (inevitably) to witness their relationships and problems.

The second stage of fieldwork, when I was able to reside in Burano, was dedicated to getting to know a wider cross-section of the population and visiting them in their homes. The house in which I lived overlooked a large grassy field, *Corte Comare*,[5] which offered a lively, active and fairly representative picture of life in the island. Removed from its main routes of circulation, the *corte* is customarily used by its inhabitants both for practical activities, such as washing, hanging laundry out to dry, polishing, grilling, and repairing fishing nets and tools, and for social or recreational ones, namely friendly interchanges with neighbours, boys' football games, women knitting or making lace together.

Taking account of Buranelli's keen interest in the most complex and extended networks of relations, both within and outside the island, I endeavoured to maintain contacts in Venice and in particular with persons working in relevant institutions such as municipal offices, schools, or medical centres. My fieldwork, then, was conducted as if within three concentric circles; first, that of my host family, secondly, the wider circle of the island community, where informants were found as far and wide as was possible, and thirdly within municipal Venice, where I followed up contacts on the basis of some relationship with people in the island. Eventually, to reconstruct the moves and the kinship connections of some of my informants, I conducted further fieldwork in Tre Porti and Sant'Erasmo – both very different lagunar environments, where the landscape is dominated by tidy vineyards and cultivated fields.

In Burano, first approaches, especially when visiting people in their homes as distinct from talking to them in the streets or *corti,* were usually formal, and proper introductions were always an advantage – a matter in which I was greatly helped by my hostess, who thought my work very worthwhile, and even amusing. Discussion of general topics, such as fishing, lacemaking, and the upbringing and education of children, was always easy and welcome, while talking of personal finances, politics and, in some instances, organised religion could rapidly come up against peoples' reserve. They would either become suspicious and defensive or, sometimes, give way to impassioned polemical utterances. In particular, research into housing, which brought me into contact with a good number of people, especially some of the elderly and disadvantaged, often led to discussion of kinship and family relations, as well as of Buranelli's preferred life-styles and their aesthetic and moral leanings.[6]

Burano's Population and the Politics of Housing

According to the census of the year that preceded my fieldwork, Burano had a population of 5,208. The number of families was 1,685. The island had a total of 1,587 dwellings, of which 67 were unoccupied, either through dereliction or because they had been purchased as holiday homes, or else kept closed by Buranelli who had taken up residence elsewhere. There were 165 more families than there were occupied houses, and house-sharing, as well as the decay of a large part of Burano's buildings, were among the islanders' most pressing concerns. I shall return to a discussion of these figures below. Before that, I shall describe the island's homes and briefly look at past proposals meant to resolve its housing problems, in the context of ongoing political debates.

Burano's homes have often been described as a typical example of 'minor', or 'spontaneous', architecture – a kind of settlement that only recently has attracted expert attention and has been redefined as 'vernacular architecture' instead of sheer 'building' (Trincanato 1948 and 1954; Goy 1989). Writers have compared the island's cottages to those of the city's peripheral working-class areas, such as, for example, Castello. However, due to Burano's increasing isolation and impoverishment from the late eighteenth to the mid-twentieth century, these cottages were characterised by even greater poverty and decay than those of the city's poorest neighbourhoods.

In their main structural features Burano's homes greatly resemble their English equivalent, that of small terraced houses and cottages. There are, however, important differences, particularly regarding the use of outside space. From the street, Burano's terraced houses never present a monotonously regular pattern in the way some English terraces do, partly because, due to the meandering nature of canals, the rows of houses are never very long and a gap is usually allowed for

Table 1.1 Population of Burano according to the 1981 census.

	0 to 14	15 to 59	59 and over	Total
Females	546	1,587	445	2,578
Males	597	1,705	328	2,630
Total	1,143	3,292	773	5,208

a narrow street, or *calle,* every four or five doors. Front doors give directly onto the street and front gardens are very rare indeed, although some homes do have a small walled garden or courtyard at their back.

A view from above shows an even greater contrast with the regularity of English terraces, since, through the building of new rooms and living areas at their back or sides and through the erection of roof terraces (*altane*), the house-tops intersect and overlap in the most interesting and complex way, so that the newcomer who wanders around in the back streets is often surprised by the obvious ingenuity employed in coping with crowding with some degree of dignity and imagination. Seen from the outside, most of Burano's houses do not appear to be very much wider than five to six, or at the most, eight metres. In the laconic description of one of the island's municipal employees, 'a traditional house mainly consists of one or two rooms on the ground floor joined to the bedrooms above by a narrow wooden staircase. But families which were really prosperous and refined used to have an attic where goods would be stored.'

While extensions have been built and rooms added wherever a space was available, many homes still have no more than two or three rooms, one above another. In some instances, however, their narrow fronts are deceptive, and as the front door opens, one is surprised to see a number of rooms, gaily decorated and taking their light from their back windows or from tiny back yards. In the simplest type of house the front door leads directly into the kitchen/living-room – usually roughly square or rectangular, with an old fireplace, now altered to form a tiled recess where an electric or gas cooker has taken the place of the old open fire.

Differences between those homes that have been modernised and those that have not been altered for the last forty or fifty years can be seen at a glance; on warm summer days, doors are left open and door curtains are drawn to let in the fresh air, so that even the most casual passer-by can take a look at a variety of kitchen or living-room arrangements. Given the minute size of rooms, furnishings and ornamental objects stand out so prominently that they almost 'speak' to the visitor with great immediacy and plainness. In the 1980s, older homes generally had free-standing wooden sideboards with glass-fronted cabinets in a fairly common 1930s style, which must have drifted in on the wake of Art Deco, while the more up-to-date frequently had adopted American or northern European style 'unit' furniture, and parlours were often merged with kitchens. In cottages that had never had the luxury of a parlour, kitchens were furnished so as to double-up as living-rooms – a type of arrangement then considered more decorous. In the late 1970s, however, there had been a return to a more intimate rustic style of free-standing sideboards with gently moulded edges, rounded corners and carvings on the backs of chairs that clearly recalled the country and hillside cottage.

Changes in taste and fashion were then readily registered in the choice of furnishings, and 'modern' still connoted prosperity and even distinction. But the most important differences between the houses that had not had any improvements made since the 1930s and 1940s and those which had been restored are functional ones: while the latter had kitchen appliances and bathroom facilities comparable to those we find in any English or American home, the more humble homes in Burano still had no bathrooms or lavatories, although most of them did have cold running water.

Overcrowding has always been one of Burano's most intractable problems. As many of my informants remembered, two-bedroom cottages would sometimes be lived in by about twelve to sixteen people ranging through three generations, but, while in earlier years discomforts were suffered in resigned silence and shame, or with little protest and little hope of finding solutions, since the 1960s and 1970s the problem of housing was felt with particular keenness. A lack of adequate living space and of the most basic facilities were clearly contrary to all the values and expectations people had absorbed through the schools and the media. As a result, awareness that their homes still did not reach modern standards, and the desire to claim for themselves what was thought to be everyone's right, often led many Buranelli (in that respect, not unlike other Venetians) to describe in vivid and dramatic terms how, in the past, they were extremely deprived of living space and to complain that as yet their condition had seen very little or no improvement.

Indeed, in Burano the question of housing – one of the political issues which appealed most strongly to popular sentiment throughout the 1960s and 1970s has a long history. Political debates, very prominent in the national media, therefore found strong echoes in the memories of my informants, for whom poor living conditions provided a rich source of imagery in the construction of a collective memory. Images of cold, humid, crowded and unhygienic cottages were frequent and effective settings for older peoples' accounts of their early years. Moreover, given that since the 1960s and 1970s standards of life in Burano had radically changed for the better, and that new differences and new hierarchies had been, and were being, created, homes were the most important and concrete symbolic objects in the islanders' constant comparisons between the past and a more favoured present, or between those who had failed and those who had succeeded in making the difficult transition between ignominious poverty and a hard-won dignity and comfort. Living in a house that had been restored and well-appointed with new furnishings, as well as the possession of domestic appliances and television sets, was the main marker of economic well-being. Indeed, I found very little or no trace of nostalgia for a homely past, as exists, for example, in England, where symbols and testimonies of the past, such as humble Victorian furniture and bric-a-brac, are diligently kept and actively sought and valued.

For those who had not been able to purchase their house and who had to live in rented accommodation, another factor in the difficult housing situation was that of the so-called *equo canone*, or fair rent law. This was mainly conceived to freeze rents, to prevent speculative house ownership, and encourage families to buy their own homes: landlords could not evict tenants unless they needed premises for their own occupation, and they could only raise rents by fixed amounts, computed on the basis of initial contracts, which sometimes went back to prewar price levels. These principles were certainly sound, but the result had been to render it financially impossible for landlords to improve, or even simply maintain, their properties, given that rents did not keep up with rates of inflation. A second consequence of the law had been the tendency on the part of many a disgruntled landlord to sell houses to people who, as future owner-occupiers, could legally evict those unfortunate tenants who were not themselves able to buy their homes. It was then ironical that when Buranelli's improved economic circumstances would have allowed them to live in more satisfactory conditions than they had been able to enjoy in the past, given the island's type of building and its state of decay, many people still could not find homes commensurate with their expectations.

While debates on industrial development, employment, transport, conservation, and, not least, the need to create barriers against high tides were followed with interest and were often discussed by Buranelli, both at political meetings and, as I learnt on my frequent boat journeys to and from Venice, in casual conversations, housing was undoubtedly the uppermost of many peoples' concerns and was one of the areas in which dwellers in the periphery shared in the problems of those who lived in the historical centre. However, as well as similarities, the situation in Burano bore significant differences from that of the city. On the one hand, a larger proportion of islanders than of Venetians actually owned their homes. On the other hand, many houses were in even worse condition than in Venice, because of their long-standing neglect and their greater exposure to high tides.

Furthermore, while in Venice relatively few people lived in ground-floor accommodation, in Burano this is the usual and preferred form of habitation. one deeply tied to Buranelli's sense of community and love of active informal interaction. Although in the nineteenth century health workers tried to draw the public's attention to the potential damage to health and the general discomfort of life in the ground floors of Venice's impoverished marginal areas and lagoon islands, remedies were always inadequate, and most of the people concerned too poor even to think of moving. Only since the 1960s, when, following Italy's national trend, the islanders started to gain a degree of economic well-being, did numerous people begin to seek new jobs and better accommodation elsewhere, often in modern high-rise blocks in the outskirts of Mestre or the Cavallino peninsula.[7] However, decisions to move were always the cause of great doubts and anxiety. As one of my most willing and alert interlocutors, Mario, then politically active, assured me:

> Buranelli only leave the island because it is impossible for families to find homes adequate to their needs ... The main point about emigration from Burano is that it is nothing like the migration from the south to the north of Italy, to find work. Nor is it

16

migration from the country to the city of people who move not to remain alone and isolated. With us it is quite different. When people leave Burano, they do not do so to find a job, nor do they leave for fear of being 'at the margins'... We have no desire to become numbers in a large metropolis; we are happy here. Even some foreigners would like to live in our islands, because here we have peace and tranquillity, and because of 'that humanising phenomenon' which is always alive.

In his view, then, the fact that most of the people who moved remained within the Venice area was clear proof that they had done so because of their need for a new home and not a search for a job or merely the love of change. The culture shock they experienced was relatively gentle compared with that of southern labourers coming to the industrialised north. No deep cultural dissimilarity, or language barrier, beyond that of a different accent, awaited them when they moved, while a well-organised transport network made it possible for them to reach their families and friends from their new neighbourhoods within one or two hours. All the same, feelings of strangeness and disorientation were sometimes very strong, because they mostly derived from uneasiness about new environments, and, in particular, noise levels, air pollution and the presence of motor-traffic.

As soon as they can, and on all holidays, they come back to see their families and friends. Many of them still grieve over the time when they had to leave: they tell us about their difficulties and maladjustments in their new environments. They still have the habits and the mentality of islanders. For older people a new residence may become a prison, because outside Burano everything is profoundly different and because they have no friends: they cannot any more sit by their front door to chat with their neighbour because their dialect and their simplicity sound strange. Some have even lost their lives because they were unable to interpret the changing colours of street lights.

Above all, while they certainly missed the familiar spaces of the streets near their homes, or their friendly main street and piazza, what they longed for the most was having a boat near the front door, and their lagoon where they could go fishing and rowing.

Not everyone shared the views so effectively expressed by my informant, since opinions on the merits and demerits of moving to inland areas were in fact widely differing. As one of his opponents maintained, he made no allowance for the fact that, especially in the 1950s and 1960s, when for many people Burano was inescapably associated with underdevelopment and poverty, cultural as well as economic and practical factors were powerful motivations for leaving the island, and with it the past, behind. At the time of my fieldwork, despite their changed circumstances, some people still held negative views of Burano, while a few enthusiasts for modernity genuinely preferred to live in the new world of factories, anonymity and cars. Moreover, with the spread of higher education, some parents said they could foresee that the island would have had little to offer their children in the way of employment commensurate with their qualifications and skills.

Mario's statements can best be understood in the context of Italian politics and of debates on Venice's environmental problems. His views on the right of all people to a decent home were in keeping with those of the city's centre-left politicians, and his aim was simply to persuade the housing authorities to provide for the needs of the islanders. At the same time, he, however, implied that relations between Venice and Burano should no longer be dictated from Venice, and that, while in the past Burano, like other estuary islands, had been thoroughly neglected, its inhabitants should start making firm demands of the larger political unit of which they were integral parts (Petris and Padoan 1978; Putnam 1993: 72–3).

To say that the planners and policy-makers of earlier years had been unconcerned with Burano would not be correct, on the contrary, the island was given a great deal of attention, especially in the General Plan of 1962. The nature and tone of that proposal, however, had caused a great deal of resentment, in particular, Buranelli found it offensive that its writers should have defined as their main objective 'the safeguard of the picturesque aspect of the island' and suggested that, to solve its housing problem, 'two thousand inhabitants should be encouraged to move, and be resettled in the areas of Murano, Cavallino, and Sacca Fisola, so that the remaining population should be guaranteed economic self-sufficiency, and an ideal housing solution' (Scano 1985: 328).

However, since that time attitudes of city architects to the population had fundamentally changed, so that the plans presented in 1978/79 (when over 1,000 people had already left since 1962) were based on the view that the population should, instead, remain stable and that economic accommodation should be built on the island of Mazzorbo where land was available. Such radical change, from a concern with Burano as a picturesque fragment of a long-lost archaic Venice to concern with its population and their continuity, clearly reflects new political tendencies and sensitivities developed over years of social confrontation and political debate. Indeed, by the mid- to late 1960s, not only policy makers, but many Buranelli too, had changed their views about the desirability of leaving the island. For, while in the 1950s and 1960s there was a tendency to consider with a degree of fatalism and resignation that, since many of Burano's homes were irremediably ruined and the cost of restoration far in excess of their worth, moving was the only alternative, by the late 1960s that attitude was partly reversed in favour of improvements to the island's housing. In particular after 1968, disenchantment with the world of industrial development and fear of becoming isolated in a 'lonely crowd' situation (one generally learnt through the media, rather than personally experienced) many people expressed a clear preference for remaining or returning to reside in Burano, even while commuting to work.

By a considerable reversal – and a remarkable sign of change – during the 1970s Burano, like other peripheral areas, began to be treated by planners as a valued opportunity. Venice's head of Urban Planning was then a professor of architecture, Edoardo Salzano, an active member of the intellectual left, in whose view restoration of peripheral areas needed urgent attention. Like other somewhat doctrinaire Marxists, Salzano analysed environmental problems due to industrial development at the internal margins of the lagoon wholly in terms of class struggle. The decay of Venice, he writes, 'is ultimately due to the fact that organisation of the natural and historical environment... has been governed solely by the search for the highest private profit on the part of the strongest propertied groups'. Burano too had undergone a severe 'territorial imbalance'; it was entirely bypassed by the capitalist processes that changed the structure and the lines of communication of the lagoon, yet its isolation and underdevelopment could be turned to advantage. Thanks to its geographical position, its future could be planned independently of the thorny problem of Marghera's industries. Issues linked with the global organisation of the lagoon could therefore be left open and undetermined, while attention would be given to those problems which, left unresolved, might have caused the demographic fall to reach a point of no return (ibid.: 36).

Because it had been so long neglected, Burano would be granted priority, and not only because its needs were of the 'utmost urgency', but also because it offered a perfect opportunity to expound social principles and approaches to planning which had been slowly elaborated over ten years of frustrating debate. The General Plan for Burano, Mazzorbo and Torcello would be 'an experiment with a new way of organising a territory' which could eventually have been extended to other areas as well' (Salzano 1978: 18–22). For example, the complexities and bureaucratic inefficiencies of Italian planning procedures could be overcome, given that plans would cover the island in its entirety and would be sufficiently detailed to avoid the need for further redrawing. Indeed, the problem of reconciling conservation with renewal was not as severe as in the historical centre, for, 'given the clear and simple settlement structure of Burano, its lack of monumental architecture... and the homogeneity both of the conditions of decay and of the culture, economy, and ways of life of the population' [sic], clearly there was no need to delimit areas with building types sufficiently coherent to allow unitary plans or to contend with problems due to the presence of different styles, or the contrast between palace and humble home often found in Venice.

In the political atmosphere of the 1960s, it had become an established principle that planning should be conducted through free consultation with the people concerned. As a first step, about thirty young architects were temporarily recruited under a youth employment scheme (law 285) to carry out a survey of almost the whole of Burano. An office was set up in the premises of the *consiglio di quartiere,* where people were encouraged to go and discuss their housing problems with the city architects.

A most interesting product of the survey, in addition to drawings of most of Burano's buildings, was a report with statistical data on about two-thirds of the total population and total number of houses – a sufficiently large sample to represent the reality of the island and of its different neighbourhoods.[8] From this survey we learn that at the end of 1981 Burano's lived-in houses were 1,520 for a population of 5,208. Most households had three or four members, about 20 percent had two members, while over 11 percent (105) had only one member. Very few households had five or more members.

From an anthropological viewpoint, partly the value of the report are the city architects' assumptions concerning family, kinship and residence, which, given their delicate role as interpreters of peoples' needs, are in themselves a key aspect of planned social change. It is, for example, of interest that *nucleo familiare* is used for 'household', while *famiglia* generally designates the nuclear family, but sometimes implicitly refers to three-generation units. Indeed one of the problems considered to require an urgent solution was that of house-sharing (*coabitazione*) by more than one nuclear family. In most instances, 73 percent, houses were shared by parents and married children.

A surprisingly large proportion of houses (70.5 percent) were owner-occupied, while only 29.5 percent were rented, and it was the latter which were usually the smallest, with an average of three rooms, while owner-occupied dwellings had an average of five rooms. Houses with four rooms were, then, the most scarce and most needed, while two- or three-roomed houses were available in larger number than were families with two or three members.

No less than 40.8 percent of the families (381) in the statistical sample were then overcrowded and failed to reach the minimum national ratio of one person: one room. Conversely, 26.9 percent (251) of families lived in houses larger than was considered necessary. To solve the problem completely, would have required 150 houses, with a total of 575 rooms. However, it was reckoned that 495 rooms were in fact in excess in relatively large houses inhabited by only one or two people, and were therefore deemed to be 'under-utilised'. (According to national planning regulations, housing standards are based on a ratio of one person: one room. A room must be at least 1.30 metres above sea-level, and ceilings must not be less than 2.20 metres from the floor.)

While the need for larger homes was met more adequately in San Martino Destro, San Martino Sinistro and San Mauro, where the original buildings have undergone the greatest transformations, Terranova and Giudecca, where we find a predominance of minute cottages, generally presented the worst conditions, with families of four or five members living in two or three rooms. One of the foremost problems was that of inadequate sanitation. For example, in Terranova, always the most disadvantaged area, only 52 percent of homes had a bathroom and over 13 percent had no indoor lavatory.

The most intractable problem, however, was that of ground floors which did not reach a sufficient height above sea-level to be secure from high tides. The situation of their inhabitants is described as 'dramatic', because tides cause humidity, they damage interiors, and generally bring about very poor living conditions. Most homes had at least one living room on the ground floor, and, where families were crowded, such rooms were frequently used as bedrooms. Only 39 percent of them reached the minimum standard height of 1.30 metres above sea-level, but, in most instances, raising the floor level was not allowed, because rooms would then have failed to reach the prescribed height of 2.20 metres, by which they are classified as inhabitable.

Proposals for Restoration

The main 'Objectives and Criteria' stated in the city architects' proposals for Burano are 'to reach the best use of the existing building patrimony and relate it to the social demand and the character of the settlement'. Coherence of building structure and function would be based on the number of floors, height above sea-level and ceiling height, as well as socio-economic, historical and typological givens.

According to the writers, crowding could have been reduced through the redistribution of inhabitants and by answering specific needs, such as those of children and the old. The main proposal, however, was that new housing for six or seven hundred people should be built in Mazzorbo. As well as helping that island, which was too sparsely inhabited to carry the cost of adequate public services, such development would have reduced the housing pressure in Burano, it would have made it possible for those whose homes were being restored to be temporarily accommodated and it might have led to some redistribution of families according to size.

Rather surprisingly – especially from the point of view of those observers who often comment on the lively animation of Burano's high street and piazza and the large extent to which life seems to be led outdoors so that all back streets and courtyards appear to be enjoyed collectively – the city architects found that the island lacked adequate public spaces and premises for meetings. In particular it was felt that older people, especially men, badly needed an indoor meeting place – a need which was felt very strongly since a noticeable increase in the number of day-tourists pushed up the prices of coffee and drinks, and the old and retired felt they had lost their dominance in public houses and bars. Moreover, a public day-nursery was desirable, while recreation facilities were thought to be entirely lacking. A bowling area could have been created in a ground-floor space then used as a storehouse, and a large hut, which was once Burano's old puppet theatre, could again be used for performances. Footpaths and routes that had been closed and private gardens that were under-utilised should have been opened to the public. Such changes, however, would have had to be made without altering the character of Burano's historic centre.

In a general way most Buranelli agreed that the city architects' proposals would have brought some improvement and that the plans were genuinely based on an understanding of peoples' needs and on attitudes of greater social responsibility than had been shown in earlier years. Despite their strong egalitarian intentions, however, as some informants were quick to point out, the architects' proposals revealed a great deal of rigidity in respect to the population, and a naive approach to family life. For example, numerous people observed that the proposed redistribution of dwellings according to family size made no allowances at all for changes related to domestic cycles, for, even if we leave aside dramatic changes in the number of household members due to separation, premature death, or marital breakdown, family composition is usually altered with the marriage of children. Couples would then have had to move house after twenty-five years or so, but, while some might have been willing or even eager to do so, to force or unduly pressure people to move at the marriage of children or onset of widowhood would have represented a form of overbearing authority most unwelcome to people as independent-thinking as Burano's inhabitants.

In so far as it purported to represent an exercise in social as well as architectural planning, the report seemed to have been based on rigid ideological premises, so that the new standards and ways of life it proposed were in fact considerably different from what inhabitants viewed as a customary part of life. This emerged in particular from observations about house-sharing. According to the survey, 12.7 percent, that is, 148 of Burano's families, lived in shared accommodation, but in most instances of house-sharing (99 out of a total number of 148 families) one, and in some instances both, of the parties involved was, in fact, a one-person 'family' – usually a widowed parent or older relative over 65 years of age.

The problem of house-sharing thus usually concerned two generations, that is, parents and children, or, as was more usual, one parent and married children, a residential pattern which until recently would not have been considered in any way anomalous or sufficiently characteristic of underdevelopment to attract public sympathy and communal intervention. The city architects' preference for 'nuclear' family residence, thus appeared a very doctrinaire one, since, although it is certainly consistent with local tradition, and it is one that was always present in the popular mind and frequently iterated in folklore, in the past it often remained an almost unattainable goal or one to be achieved gradually over the first few years of marriage.

In this respect policy-makers in the 1970s and 1980s were true interpreters of people's desires, but the fact that the word 'family' sometimes designates exclusively the nuclear family, but at other times (as is more consistent with current Venetian and Italian usage) is used loosely to mean 'household', 'extended family', or 'family of origin', indicates that supposedly obsolete mental attitudes and conceptions of 'family' were in fact still strongly present in the minds and in the language habits of planners and operators of social change. Most striking of all, however, was the writers' cryptic value-judgment on the situation of those young families who had to live in the same home with either one or two grandparents: 'this datum is not entirely negative if it is viewed from a social viewpoint and in respect to the handing down of cultural traditions to new generations, but it is contrary to present living models'.

One very rewarding, though incidental, aspect of that phase of my research was that it offered ample opportunities to observe relations between the islanders and the young architects as employees of the municipality. As some of them reported, their reception in the homes was rather mixed and unpredictable, as some people regarded them as instruments of those city authorities they had traditionally defied or held in some suspicion. Ironically, it was often the people who needed their help the least who were the most willing and ready to throw their doors open, while those whose homes were in need of updating were sometimes rather embarrassed and reluctant to make their shortcomings known. Lack of toilet and bathroom facilities was felt as a humiliation and, in a number of instances, doors were abruptly closed in the faces of surveyors, with the question: 'Say, young man, have I come to your house to make such inquiries?' Rejection was not only motivated by a sense of shame and of injured privacy, but in a number of instances it was also due to people's awareness that extensions they had built and improvements they had made to their homes were contrary to the city's stringent building regulations, and could make them liable to be fined or even ordered to demolish the recent additions to their living quarters.

The way many Buranelli felt was that, while they undoubtedly had achieved some economic well-being, they had done so on their own and certainly not thanks to Venetian or national help. Could such detailed and inquisitive interest in their homes on the part of the municipality really do any good? A large part of those improvements they had brought to their houses had been, so to say, 'spontaneous', but would they now be found to be 'irregular'? Their fear that they might be penalised by the city authorities for the little progress they had made on their own was then possibly greater than the confidence that housing problems would ultimately be resolved to any satisfactory degree thanks to communal intervention. Despite misgivings and a few refusals to cooperate, most people did open their doors to the young architects. 'After all', several women commented, 'poor things, they too had to eat: if it weren't for the survey, they would be unemployed!'

A large part of the island was drawn, charted and photographed. When a plan was eventually elaborated and approved, it too was greeted with a certain amount of skepticism: the new neighbourhood in Mazzorbo, it was felt, was definitely going to be too compact and the design was thought to be out of keeping with its environment. But dislike of the new housing development, expressed mostly while it was only half built, may have been due partly to the fact that the houses were not yet painted, for, as I note elsewhere, Buranelli regard any colourless building as insipid and tasteless (for example, schoolchildren on a trip to Perugia kept remarking on how beautiful its battlements and buildings would be, if only they could be decorated in different colours). Other negative comments concerned the manner in which building works had been organised and conducted. Because the development was financed with public money, too many firms were called in, yet builders were very slow and the whole operation had repeatedly come to a standstill, while the waste was enormous. It also seemed to have been a foregone conclusion among some of Burano's inhabitants that the new neighbourhood would largely have been lived in by those undesirable or inept families who had been unable to establish a home without social assistance.

Despite the housing crisis, then, turning to public agencies for help was still associated with poverty and, although not universally, still carried a slight stigma as a surrender of independence and an admission of personal inadequacy. The greatest doubts concerned the manner of allocation, which, it was feared, would inevitably have taken place along patronage lines rather than on the basis of a correct evaluation of needs and priorities. Nevertheless, despite some initial diffidence on the part of prospective occupants, and criticism of departures from a traditional use of colour and design, the housing has actually proved very successful (figure 1.18).[9]

However, some Buranelli complained to me about the city architects' poor perception of social and family structures. In particular, they resented a description of their society as 'homogeneous, both in the conditions of decay, and in the culture, economy and way of life of its population', as well as the definition of Burano as a 'suburb' or a 'dormitory area', because, they explained, a suburb would have been developed later and have been culturally dependent on the city centre. On the contrary, Burano, which may have been *prior* to Venice, has its own ancient culture and its own traditions.

Notes

1. 'The reader,' Ruskin writes, 'may perhaps have felt some pain in the contrast between this faithful view of the site of the Venetian Throne and the romantic conception of it which we ordinarily form'(1887, vol. II: 7).
2. The total territory of the Venice commune is 457.47 km². Of these, 189.84, that is, approximately 40 percent are dry, and 60 percent are covered in water. Inland areas are predominantly dry, while the ratio of land to water in the historical centre is on average 13.2 percent of the total. Burano's area is 50.53 square kilometers, of these only 1.1 percent, 0.56 Km² are dry land (Comune di Venezia 1989: V).
3. This is less true since the early 1980s, because, with the development of mass tourism, many visitors use the city transport and Buranelli often have to fight for a seat and look for their friends through crowds of 'foreigners'. .
4. Ciardi, Rossi, De Marchi and other post-impressionist painters used to spend some time in Burano in the Autumn. Semeghini actually had a home on the island. They regarded Burano as their Brittany, a place where they could escape the pressures of exclusive international art circles in Venice. Numerous Buranelli also paint in their spare time, and a few have become successful professional painters. To acknowdge what is now a well-established tradition, a prize, *Premio Burano*, is awarded every year at an exhibition held in the *Consiglio di Quartiere*.
5. A *corte* is a square, or common, surrounded by houses.
6. Efforts at finding solutions for long-standing inadequacies, which are now viewed within the general context of the 'problem of Venice', provided a first example of interactions between islanders and city officials.

7. The number of people leaving Burano was highest in the years 1951, 1959, 1960 and 1961–63. At the close of 1951, the population had gone down by 174, including those who had died; 212 more had either died or left the island in 1959–60; and 612 between 1960 and 1963.

8. *Comune di Venezia. Assessorato All'Urbanistica. Piano Particolareggiato del Centro Storico di Burano.* Relazione (not dated).

9. Never having gone through a 'modern' phase, the islands provided an ideal venue for a housing estate with post-modern features.

Figure 1.1 Burano from the air. Via Baldassare Galuppi.

Figure 1.2 Ground plan of Burano. Venice. Asserato all'urbanistica.

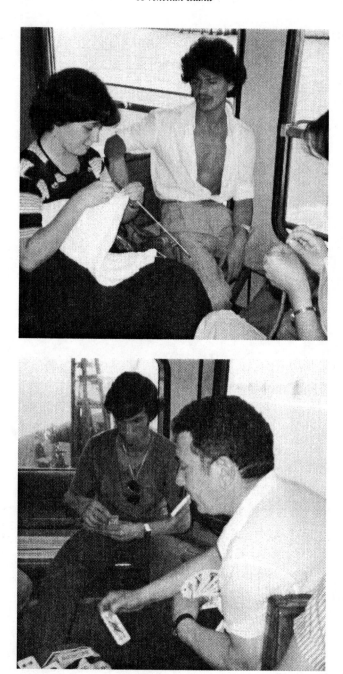

Figures 1.3 and 1.4 The boat back from Venice. Regular commuters settle down to some handiwork, conversation or card games.

Figure 1.5 Children playing in *Corte Comare.*

Figure 1.6 Gendered games. By the age of twelve or thirteen boys play with boys, girls with girls.

Figure 1.7 Life in the streets. Although most people own electrical appliances, domestic tasks are sometimes carried out in the alley, courtyard or street outside the house.

Figure 1.8 Grilling fish for Sunday lunch.

Figure 1.9 Helping grandmother.

Figure 1.10 Children at work. Copper implements must be regularly polished to a very high shine.

Figure 1.11 A well-earned rest.

Figure 1.12 Old friends exchange news and gossip.

Figure 1.13 The nursery school.

Figure 1.14 Lace stall.

Figure 1.15 The market: closing time

Figure 1.16 Evening at the café. Buranelli love to sing. They have a distinguished choir with a repertoire that includes traditional folk songs, as well as opera.

Figure 1.17 Boats at dusk.

Figure 1.18 New housing at Mazzorbo.

2
A SENSE OF HISTORY

Two questions…were set up against one another. – 'How did the past lead to the present?', and 'How does the present create the past?'
                                    Tonkin, McDonald and Chapman, *History and Ethnicity*

When, in the early days of my fieldwork, I explained my interest in local history, several people told me that, before settling in Burano, their community had lived in a different island, Buranello, Boreano, or Burano da Mar. It was not very far from their present location, but, as it was exposed to the winds and the sea, it was destroyed by tempests and disappeared under the waters. They, therefore, arrived in Burano like shipwrecks, refugees twice over, first from inland Altino, fleeing from barbarian invaders, then from the elements.

The legend is also reported in a few passages by those rare Venetian historians who gave Burano more than a passing mention. As Coronelli writes,

> Some people maintain that the population of Burano may have been increased by refugees from Buranello, a small island about five miles away in the direction of Sirocco; vestiges of its ruins can still be seen under the water, when the sea is clear. (1696: 33–34)

Also Flaminio Corner, almost echoing Coronelli, reports that, according to tradition, Burano would have been settled by inhabitants of a nearby island when that proved so 'badly exposed and gravely threatened by the impetuous currents of the sea that they feared they might all be submerged'. The move would have occurred in 959, and Burano would then have been known as *Burano Nuovo*. There, according to Corner, the settlers would have built the church of San Martino, named after the Bishop of Tours, but the church was only consecrated by the Bishop of Torcello at a much later date, in 1630.

Both writers, however, introduce their mention of an Ur-Burano with a degree of skepticism, describing it as a vague and obscure folk memory. Like them, at the time of my fieldwork not all Buranelli gave the legend much credence, although they did often mention and discuss it. Several fishermen claimed that when the weather conditions are favourable, the sky is clear and the water transparent, the ruins of Buranello can be seen at the bottom of the lagoon, but others assured me that they thought such an island had never existed. Are today's Buranelli echoing the words of seventeenth- and eighteenth- century historians? Or did the historians, like latter-day anthropologists, take note of the Buranelli's ancestors' narratives?

The legend may well be based on a real event, since, after all, several islands are known to have given way to erosion and disappeared after their inhabitants abandoned them to escape from malarial infection and seek safer and healthier environments. Archaeological investigation and photography may soon provide an answer, meanwhile, the legendary existence of an earlier Burano remains a subject of interest and fascination. Indeed, both the story and its aura of uncertainty were typical of Buranelli's feelings about their island's history at the time of my fieldwork.

Not long after my arrival members of the library committee had set themselves the task of putting together an archive that would document the island's history – an operation of recovery then widespread throughout the country and partly motivated by a fear that intense social change might bring about a loss of memory and identity. A poster with the heading 'Burano and its Vicinity: ... a history puzzle' (figure 2.1) affixed on the walls and circulated in the schools, showed an early photographic view of Burano, fragmented by the superimposed lines of a puzzle, over the following words:

> In the context of our initiative to create a local historical archive, Burano's library is promoting a collection of historical and documentary materials concerning the environment, and the customs of Burano and its lagoon. The population and cultural associations are asked to make available materials they may possess: old photos, especially family and wedding ones, postcards, manuscripts, sound recordings of the dialect, or any other materials that may illustrate the life-styles, customs, political life, the world of labour and the social and economic changes that have taken place in our area.

The documents would be copied and then returned. The project, comparable to those of 'History Workshop', was to record as much of Burano's history as could be reconstructed from memory and from material objects, that is, at the most four generations.[1] Speculation about the legendary existence of an earlier Burano and the islanders' plan to record a far more recent past, were both signs of a need to have, or to make, history at the two extremes of the island's social time, its beginnings in a remote past and its present, or indeed a future when, it was hoped, the archival collection would have been of value to new generations.

A desire for a 'beginning' was also due to the fact of being in a culture area in which foundation stories are ever present (Fortini-Brown 1996), but where

Burano's settlement is generally treated as a detail in narratives about its more illustrious neighbour, Torcello. Medieval chronicles begin with accounts of the flight from Altino of the Christian faithful, under the pressure of barbarian invaders. The *Chronicon Gradense* explains that the island was settled by people who had previously lived near the northern gate of Altino, Borea, hence its name, *Vicum Burianum.*[2]

The idea of a common origin, however, is sometimes denied by the Torcellani, who claim that only they are the true descendants of the Altinati, and who argue that Buranelli could not possibly have come down from a population of cultivators, since they were not in any way like peasants (*contadini*). To add to the mutual stereotyping, it is sometimes suggested that the population of Torcello must have been Altino's aristocracy, while Burano's inhabitants may have been mainly plebeians who followed their superiors in their flight and joined a community of humble fishermen, settled there since time immemorial. Indeed, inhabitants of neighbouring islands always emphasise the Buranelli's difference, echoing vague conjectures about their supposed oriental roots, or suggestions that Burano, as well as areas of the coast near Altino, may have been Greek colonies settled since pre-Roman times, while both claims are vigorously denied by those partial to the idea of Roman origins.[3]

Partly the reason for such uncertainties is the fact that throughout that early period the cultural and economic centre of the northern lagoon was Torcello, first the seat of government and residence of the tribunes in Roman times, then seat of the Bishop and *podestà* under Venetian rule. When Torcello lost its dominant position, its population decimated by malaria and its churches and convents abandoned and ruined (see below), Burano became the administrative centre of the small estuary. By that time, however, all political and economic activities had moved to the Rialto area, while the islands of the Northern lagoon had lost all their power and prestige. Thanks to the presence of its late Roman and Byzantine monuments and ruins, valued by humanists and later by romantics, Torcello, which Ruskin described as 'mother' to Venice (1867: 12), retained the prestige of antiquity, but Burano, which had no comparable architectural or historical remains, was increasingly treated as a remote and isolated area of little interest or significance.

An awareness of the overwhelming production of historical works on Venice made the silence surrounding Burano even more humbling, so that its inhabitants were faced with the need to negotiate their pride in sharing in Venice's antiquity and political history, with a firm conviction that theirs was a distinct community – one marked with vivid memories of past poverty and marginalisation. Their feelings no doubt also derived from the prestige of the written word, which, as they learnt in their early school years, and were often reminded by representatives of the dominant culture, was held superior to tradition and memory.

Having records of the past was associated with prosperity, reputation and respectability (somewhat like owning family portraits), while the lack of any such records seemed a proof of their insignificance. In contrast to inhabitants of

Venice's centre who feel oppressed by a surfeit of history, Buranelli, then, saw their absence from most official versions as evidence of a hurtful lack of recognition by historians. Several people recalled that because they had long suffered from a low level of literacy, they had felt a sense of dissatisfaction at the difficult confrontation of vernacular traditions with the national language and culture.

Despite such feelings, and despite some uncertainty about the extent to which they shared in Torcello's history, many Buranelli proudly emphasised that their village too was settled earlier than Venice. Legendary versions of Attila's arrival at Torcello (in 452 A D) were often related with a degree of skepticism, although the expression 'he is a real Attila',[4] was sometimes used to describe the most restless and unruly boys. Christian origins were often proclaimed, and legendary accounts of the miraculous landing in Burano of its patron saints were known to most people, since their memory, charmingly recorded in a large canvas in Burano's church of San Martino (figure 2.2) was also kept alive by grandparents and Sunday school teachers. Counter to those, however, as we have seen, were speculations about supposedly foreign or non-Christian elements that might have formed some strata of the population. For example, the surname Barbaro, 'barbarian', was said to have derived from the nickname of a barbarian slave who had served in the household of one of Torcello's Roman tribunes: he was eventually freed and went to live in Burano, where he handed down to his descendants the secrets and refinements of a rich cuisine, which he had learnt during his captivity.

A favoured theme was that of freedom and independence. Some people recalled with pride that, when, in the ninth century, Venice became the Lagoon's political centre, the islands continued to administer themselves in accordance with ancient municipal statutes and institutions. In particular, Torcello and Burano retained some privileges, which, according to tradition, could be traced back to Roman or pre-Roman times. A sixteenth-century actor and comedy writer, Calmo (below), described himself (perhaps in jest) as 'a descendant of good old Torcellans, all...faithful keepers of their jurisdiction.' But, although memories of political autonomy and independence still inform some of the islanders' attitudes and sometimes enter political debates, documentary evidence must have been long lost. As a historian observed, in 1441 the Venetian Senate passed a motion to ask that the people of Torcello should not continue to refer to customary laws *de quibus tamen non reperitur aliqua scriptura* (of which no record was found) and ordered the *podestà* to observe the laws and ordinances of Venice and to draw from approved local custom only when laws were lacking (Cozzi 1982: 11 but cf De Biasi 1994).

A striking feature of Buranelli's accounts was the lack of detail for almost a whole millennium of Venetian rule; the time dimension was completely flattened out, and Burano's social history viewed as one of unremitting poverty and isolation. Images of the island as one of poor 'fishermen and lacemakers' then became almost timeless, while the few better-off and successful people were, in a way, made invisible. Burano's one really famed individual, the composer Baldassare Galuppi (1706–1785), after whom are named the main street and piazza, was remembered with pride, but it was often observed that as soon as he had reached success, his life and work had become part of that of Venice – or of the world – while Burano was only mentioned as his humble birthplace.

In the latter part of the eighteenth century, when the Venetian economy was in a state of recession and disarray, and to an even greater extent after the Veneto was taken over by Austria in 1815, Burano too was severely impoverished and increasingly isolated. It was especially from that time that its image as a community of paupers became fixed – both in the minds of the islanders, and in the eyes of its various visitors, benefactors, and educators. References to Burano then began to enter the writings of the philanthropists and politicians who contributed to the organisation of its Lace School (chapter 4).

Historical interest was naturally much livelier as more recent events were touched on, and the topics most old people remembered with a great wealth of detail were the advent of fascism and the Second World War. The joining of Burano to the Venice commune in 1924, the suppression of the democratic election of mayors and the return to rule by a *podestà*, as in the days of Austrian domination, were often mentioned by the more politically conscious. Episodes that occurred during the German occupation, such as encounters with the fascist Black Brigades, executions of partisans on a nearby abandoned island and the Allied bombing of a house, were all recorded in many people's minds and were usually transmitted to the very young by their grandparents.

Questions about origins, hinted at in the heading of the poster and the puzzle image (above), but wisely left in abeyance by those who collected materials for an archive, remained a frequent theme in Buranelli's historical discourses. Like sophisticated ironists, several people then offered me their own suppositions about the circumstances that might have led to settlement in their island, or in general about their past, through self-deprecating or semi-serious myth-making, slight narratives and in-jokes (often offered to me complete with exegesis). And, while these cannot, in any sense, be described as fully-developed and articulate myths, but rather as conjectures and deliberate fantasies, they are, nonetheless, of interest as indications of attitudes and sentiments, rather than real answers to Burano's historical questioning. Stories, carnival sketches, segments of history and, in some instances, also life-histories were assembled in such a way as to leave transparently in evidence superimposed images and bricolage pieces, so that an attentive and well-informed observer could probably identify, disentangle and relocate the fragments.[5]

Answers to unresolved questions were sometimes provided in the form of imaginative tales or just-so stories, often told in facetious and ironical self-mockery. A common underlying theme was that of Buranelli's slight physical 'difference' from inhabitants of neighbouring islands. By the time of my fieldwork such differences had entirely disappeared, but some people recalled that in the past Buranelli used to be somewhat shorter and slightly darker-skinned than their neighbours, although blue eyes were not at all unusual.[6]

The most common and, in their view, most plausible explanation was that they are simply the descendants of *povera gente* (poor folk), but another possibility Buranelli sometimes contemplated was, as we have seen, that of some ill-defined, but 'different', ethnic origin from that of other Venetians. One suggestion was that Burano was originally a penal colony where Venetians relegated thieves and prisoners of war, and more than one informant said that they had an obscure certainty that the physical characteristics of Buranelli showed that they must have descended from Turks. The name of a street, *Calle degli Assassini*, or Street of the Assassins, was once explained to me by a woman in terms of a crime of passion, committed by one of her ancestors who was insanely jealous of his wife. The street name *Assassini*, a word derived from hashish, is also found in Venice, where it is usually explained as designating an area where sailors smoked and traded that drug. Both glosses lend support to some suspicion of Turkish influence or presence, but, in the woman's dramatic interpretation of the street name, a hot temper and violent defence of honour was viewed as strong evidence of 'southern', and possibly 'infidel', character traits.

The presence of a number of Jewish-seeming family names, as well as the fact that a kind of biscuit traditionally made in Burano curiously resembles some types of Passover cakes, and, above all, that one of Burano's neighbourhoods is called Giudecca, had led people to suppose that some families must be Jewish in origin. As Guiton writes, quoting a Buranello friend, 'Then there is the Jews' bridge. There were never many Jews here in the old days, but there were always some, which is why we had a Giudecca' (1977: 117). However, 'Giudecca' may not be linguistically related to 'Jew', but to the past participle of the verb ' to judge', giudicare', in Venetian *Zudegà*, which would support a view of Burano as a place of confinement or prison. The Venice Giudecca, initially called Spinalunga from its shape, would owe its present name to the fact that some of its areas were assigned to a number of Venetian families when, after a period of political exile, they were pardoned and encouraged to settle there. *Zudegà* would then refer either to the fact that such families had undergone a 'judgment' in the sense of a trial and sentence, or to the fact that properties were 'adjudicated' by the drawing of lots (Correr, G., et al. 1847, Vol. I: 493; Ravid 1977: 201–225).[7]

An imaginative version of the early history of Burano was offered to me when I asked a man why the islanders paint their houses in many different colours. His answer was:

> Because once upon a time Burano was one great isolation hospital. When there were cholera epidemics, individuals or whole families would be sent here. The Republic's sanitary officers would come and spray our houses inside-out with quick-lime, but the white-wash made them look desolate and almost sinister, so that, as soon as all

members of a household were declared free from infection, they would come out and paint their cottage in vivid colours to celebrate, and to let others know that they could join society again.

Here historical memories about the nearby island of Lazzaretto Nuovo, which was indeed an isolation hospital, are extended (or symbolically identified with?) Burano.[8] Such sporadic inventions in which bits of history were woven with fantasy could hardly be considered to form a well-developed body of myth. What is of interest, however, is that some common traits run through many of them, and those are the themes of marginality and exclusion, expressed through symbols of stigma, such as contagious illness or the fact of being at odds with the law, or through examples of embattled independence and defiant intolerance of authority.

A revealing example of this was the Buranelli's explanation of the way their local archives were lost. The first time I inquired, I was simply told that the municipal offices were set alight 'by a madman' and all archival materials had been destroyed before the flames could be put out. A much fuller account, however, was spontaneously given me a few months later: it was the winter of 1944 and people were suffering from cold and hunger. Somehow the Buranelli had managed to have a greater number of food ration cards then their population justified, but, when they were warned that (whether due to treachery or indiscretion) the Black Brigades then controlling Venice had become suspicious and would soon come with German soldiers to check Burano's municipal documents, the commune's offices were set on fire. The incident is usually related as a gesture of resistance against outside interference, and a deliberate act to destroy evidence that would have allowed their wartime rulers to impose strict controls on food rationing and possibly inflict some dreadful reprisal.

Despite a feeling of loss and the knowledge that some gaps may be very difficult to fill, Buranelli said that they knew that relevant documents must have been kept in Venice's archives, but they had not yet been fully ordered and examined. To view the history of Burano as irretrievably lost would have been a mistake, given that even on a first attempt to collect a bibliography I found that literary and archival documents are certainly in existence and remain partly unexplored (cf. De Biasio 1994).

Meanwhile, I collected a variety of writings on the northern lagoon. The result was a small number of texts of diverse styles and periods: references to Buranelli in a few sonnets, a fictitious letter and a poem in Burano's dialect by a Renaissance comic poet, Calmo, illustrate early versions of enduring stereotypes; descriptions of Burano and brief accounts of its history by a geographer and a Church historian shed some light on the island at the turn of the seventeenth century, while the writings of a late eighteenth-century scholar show the continuity of earlier representations, as well as a tendency, widespread in that period, to equate peasants and fishermen with primitives.

A debate which followed Italian unification in the 1860s, and was revived in the 1920s, concerning whether Burano should have been joined to the Venice commune illustrates important aspects of the history of relations between Venice and its island periphery. For the nineteenth century we learn much about Burano's history from writings about its lacemakers and Lace School. Contrary to a general silence on the history of women, in this case it is the women who are the main protagonists and who are the link between Burano and those parts of the world involved in the international lace trade.

Burano's Fishermen as Comic Stereotypes in Renaissance Drama

Glimpses into life in the lagoons in the sixteenth century can be found in a few texts by an actor, poet and comedy writer, Andrea Calmo (1509–1571), in whose works Burano's fishermen are represented as the Venetian equivalents of rude peasants and comic villains in Ruzante's Paduan comedies. Very little is known about Calmo's life, but passages from his works suggest that he may have been born or brought up in the northern lagoon area 'among baskets, fishing rods and nets'.[9] References to such humble origins may have been ironical, but several of his works show that he was intimately acquainted with the speech of the northern lagoon islands, indeed they are often referred to by scholars as rich sources of linguistic knowledge (Filiasi, below, and Padoan, 1982: 159–60). In his early comedies, which reflect the realities of Venice as a lively cosmopolitan sea port, he explores, exaggerates and parodies the manners and speech of its diverse and colourful population, at the same time portraying the true linguistic preferences of its merchants, craftsmen, working people, clerics and aristocracy (Devoto 1960: 70–1; Dionisotti 1968: 1; Migliorini 1984: 214–5; Sciama 1992: 361–2).

While earlier writers in the Venetian vernacular had tried to develop a genteel aulic dialect, Calmo's characters, a medley of Bergamasque porters, Greek, Slav or Istrian sailors, prostitutes, citizens, maidservants and fishermen, enacting their fortuitous and comical encounters in a crowded urban setting, each speak her or his own patois. The writer thus goes through a wide range of dialects, from those of the Paduan and Bergamasque countryside, respectively associated with servants and porters, to *Grechesco*, a lingua franca with large numbers of Cretan or Rhodian words, mainly spoken by Greek and Albanian sailors and mercenaries (or *Strathiotti*, the Venetian equivalents of Rome's *Milites Gloriosi*).[10]

In Calmo's vivid dramatisation of the meetings of such diverse people, Buranelli are characterised as behaving according to 'nature's way', *naturaleza,* and their speech and manners are contrasted with the stilted diction, and aulic dialect of Petrarchan love poetry, and, in general, all social and literary affectations. Like the best of parody, Calmo's portrayals of the islanders, in which fantasy is woven in with acute observation, successfully capture the essential traits of their behaviour and speech.

In his *Epitaphs,* a parody of the mannered and lengthy poetical epigraphs of renaissance tombstones, he records the circumstances of the lives and deaths of his characters (Calmo 1600: 54 ff.). While his parodic mode clearly leads to exaggeration, nevertheless the *Epitaphs,* rather like the *Spoon River Anthology,* show a very keen understanding of the manners and preoccupations of lagoon islanders. The inscriptions engraved on the burial stones record essential character traits, events, ambitions and desires in the lives of those who are buried, and they thus highlight motifs and concerns that friends in Burano assured me they recognise as true to island life. The main themes are those of self-worth and desire for honour and respect; despite the indignities of poverty, people universally demand to have a good burial. They also feel great pride in their descent, and they express strong male solidarity. The peer group, and, in particular, the community of men (*brigae*), are very important, but friends are sometimes treacherous.

As well as their sense of the fulfillment of life through love, satiety, companionship and the reassuring certainty of a line of descent, several of the epitaphs recount the circumstances of violent deaths, often meted out in the solitude of the marshes or among the high grasses and reeds of fishing enclosures. Indeed, several of the poems record a tendency to take revenge even for some imagined or unimportant slight, or they sadly recount the frequent occurrence of drowning or of accidents at sea. The stereotypes created by Calmo appear to have been based on a very close and intimate knowledge of his subjects, and seem in turn to have had an influence on their own self-perception and self-presentation: as if to play up to Calmo's burlesque representations of island life, at Carnival time Buranelli still impersonate some of his stock characters in laughter and self-mockery.

The History of the Northern Lagoon from the Works of Vincenzo Coronelli, and from Flaminio Corner's Ecclesiastical History

In contrast to Calmo's flamboyant parodies, Vicenzo Coronelli and Flaminio Corner [11] both describe at length the sombre life of the northern lagoon's numerous convents – undoubtedly the places where Burano's lace craft was sedulously practiced and brought to technical perfection, and where ideas about gender, sexuality and shame became deeply-rooted *habitus.*

Both writers report the folk memory according to which Burano was settled by inhabitants of another island when the latter was submerged. Coronelli's description of that event as 'popular legend' and 'common opinion' shows how the mixing and mutual influences of oral traditions and recorded history actually renders them inseparable. In antiquity, Coronelli writes, the islands were known as 'the six sestieri of Altino', a city named after a Trojan hero, built 'entirely in the style of Ravenna' by Antenor on the shores of the rivers Zero and Sile, just eight miles from Treviso. When Altino was destroyed during Attila's invasion, its people were able to escape and reach the islands from the rivers thanks to their mastery of seamanship. The nobility then established themselves in Equilio and Eraclea, while the 'plebs' mostly settled in Torcello, Burano, Mazzorbo, Costanziaco, Ammiana and Murano.

Coronelli then describes Torcello's great power, when it was the residence of the *podestà* who also governed Mazzorbo and Burano, and when, from 635, it was also the seat of the bishop, whose rich diocese owned lands on the shores of the river Sile and included no less than twelve parishes and sixteen convents of nuns. By the time of his writing, however, the population had been decimated by malaria and 'such was the unwholesomeness of its air, that, but for the convents that adorned the city and that were used as places of retirement and seclusion by highly qualified ladies, Torcello would have been entirely deserted' (1696: 33).

Religious communities, like the rest of the population, were always sensitive to the effects and echoes of war, to the intemperance of the climate and the decay of buildings, which forced them to move from convent to convent.[12] Images of a more prosperous past also characterise Coronelli's description of Mazzorbo. When it was first settled by refugees from Altino, and when, later, it had numerous country houses 'for the enjoyment and delight of the nobility', it must have been much different, but at the time of his visit, he writes:

> nothing worth notice is to be found in Mazzorbo, except piety, which flourishes in a truly exemplary manner, in four convents of nuns... all under the rule of Saint Benedict. In them there shines the Venetian nobility, who, with generous contempt for the fasts and luxuries of the world, see their daughters most religiously cloistered, attracted by the pleasant solitude and the venerable antiquity of the islands, and by the famous relics kept in their sanctuaries (Ibid.).[13]

Burano also had a relatively large religious population, which included two Benedictine convents: one, which housed forty nuns, by the quay of San Moro and the other in San Vito, close to the piazza. A third convent, Santa Maria delle Grazie, was of the order of Capuchin Servites.[14] In addition, it had no less than forty priests, an oratory and a hospice for the poor. Like Mazzorbo, however, the island presented an image of decay: the bridge that joined the two islands was thoroughly ruined 'through carelessness and time', but its population was nonetheless livelier and more numerous than elsewhere in the northern lagoon.

> The streets are sometimes too narrow. The islanders are completely dedicated to fishing and to the construction of ships, by which Burano acquires its riches and renders important services to its rulers. The women work at *punto in aria* and they weave nets. Buranelli are always in competition with the inhabitants of Torcello and Mazzorbo, and, although close to the direst poverty, they boast noble origins (Coronelli 33–34).

As we have seen, for Corner the history of Burano coincides with the foundation of its first church, San Martino, in 959, but that date, as he writes,' so far remains uncorroborated by other evidence'.

Events that occurred in the Renaissance period are naturally told with greater assurance. Of particular interest is the arrival from Noale of nuns fleeing from the violence of war in 1514 [15] and their resourceful foundation of a Benedictine convent on the ruins of the old church of San Vito (1749: 600 &ff.).With them, also numerous lay refugees must have taken shelter in Burano – one may wonder whether, and to what extent, the nuns had a role in the development of organised lacemaking among the island women.

The O-Tai-Tans of the Lagoons

A work that eventually exercised considerable influence on later scholars, and contributed most strongly to a view of Burano as an authentic repository of Venetian tradition, is Count Jacopo Filiasi's *Memorie Storiche de' Veneti Primi e Secondi* (1796). Writing at a time of intense political change and keen interest in ethnic origins, and drawing his evidence from archaeological as well as literary and linguistic sources, Filiasi attempted to fill the long gap between prehistoric, Atestine and Roman Venetia. In the words of a later historian, Filiasi tried to bring to light 'the moral character and the national and spiritual physiognomy' of the early Veneti (Cessi 1944: 389).

Although partly of the kind anthropologists would later have described as 'conjectural', Filiasi's work provided a lively picture of early insular Venice, thus enriching both popular and scholarly traditions and laying the basis for an early romantic vision of Burano as a precious survival and testimony to a primordial Venetian identity.[16] As he maintains in one of his most speculative chapters, before Romanisation, and certainly long before Attila's invasion in the fifth century and the settlement of refugees in the islands, the north-eastern corner of Italy was inhabited by a population of Asian origin, the Heneti, or *Veneti Primi*, who had originated in Babylonia and Assyria. After migrating to Paphlagonia, on the northern border of the Black Sea, they would have moved to Illyria, eventually to spread throughout the north-east of Italy and settle in the area that lies between the sea, the Friulan mountains, the river Timavo and Lake Garda, and extends to present-day Brescia, then populated by Celtic Gauls.

According to Filiasi, Asian origins are also proved by the early Veneti's cult of Belen, their worship of the Dioscuri and their periodical sacrifice of a white horse to the Thracian king Diomedes. In Abano, now a health resort where hot mud springs are used for therapeutic purposes, they observed an oracle of Gerion, and at the baths there was found an inscription to Isis, the Goddess of nature. The Veneti's usual costume also recalled Oriental dress. According to descriptions by Juvenal and Strabo, men wore a sleeved tunic over wide trousers and a hat on short curly hair, while women were veiled as in eastern countries; they did not often appear in public and they were considered extremely chaste. The favourite colour of the Veneti was deep blue, and their clothes were still that colour in the twelfth century.

The clearest marks of Asian origins, which, in Filiasi's view, were still present at the time of his writing, were the Venetians' accent and language. Although their early speech was greatly altered through mixing with the tongues of Etruscan, Umbrian and Euganean tribes, and in time it entirely gave way to Latin, it nonetheless kept its original accent, as well as some words.[17] Numerous place names and dialect words were derived from Greek words, but these were themselves formed from the languages of Assyrians and Paphlagonians. Such influence was particularly evident from the abundance of final diphthongs, which recalled the Ionic dialect, as well as the lengthening of vowels and a kind of singsong or lilt, still typical of the speech of Chioggia and Burano.[18] As Filiasi writes,

> It is not rare still to find among Buranelli fishermen some of the most illustrious and ancient surnames, i.e., those of families which are already extinct elsewhere.[19] Their dialect, and especially their accent and pronunciation, are very different from those of other islands. They drag their words and they lengthen and double their vowels in the highest degree. 'Paare', they say and 'Buraa' instead of 'pare' and 'Burà' as others do... In this the Buranelli can truly be said to be the O-tai-tans of the lagoon (1796, Vol.4: 203).[20]

According to Filiasi's theory that accents are the most enduring feature of languages, Buranelli's way of pronouncing first Greek then Latin words, 'almost singing', must have been a characteristic of a particular groups of early Veneti and of those inhabitants of Oderzo and Altino who fled to the islands from the area between the rivers Sile and Livenza under the pressure of Barbarian invasions, in the fifth century. Unlike the populations which remained in inland areas, such as, for example, Treviso, the islanders never came into contact with their barbarian conquerors and their speech remained close to its original forms. Because of their bizarre singsong, which was 'imitated by poets, like Calmo in his writing of national, that is, Venetian, comedies they have become the models for...the well-known mask of the *Gnaga* which is a parody of the Buranelli and their speech. (Ibid: 10)

Implicitly echoing debates between linguistic purists and those who wished for a more dynamic and freely created Italian national language than one strictly based on the fourteenth century Tuscan of Dante, Petrarch and Boccaccio, Filiasi shows his regional loyalty in defending the speech of Venetians against the criticism of Dante, who described it as 'so raucous that a man talking sounded like a female (*De Vulgari Eloquentia*). On the contrary, Filiasi writes, Venetians were endowed with remarkable musical sensitivity, but then, Filiasi concludes, 'Dante's judgements do not deserve to be respected just because he is Dante...he may have spoken so just to humble a little the peoples of Venetia. Possibly because in those days they valued and praised their own dialects too much' (ibid. 204).

Filiasi's defence of Venetian speech makes an interesting comment on Italy's long-standing 'language question', just before the dawn of Italian state formation. Indeed, a search for origins and for memories and symbols of early tribal groups and linguistic forms was widespread among late eighteenth- and nineteenth-century scholars as they sought to identify earlier ethnic cores that would become part of a wider Italian nation (A. Smith 1988: 61). By his appeal to a primitive root of an imagined Venetian identity, Filiasi's aim was in the first place to establish the great antiquity of the bond between people and territory, but also, and perhaps more importantly, to supersede the debate over Roman versus Greek origins of Venice and to counter too full an intellectual and linguistic subjection to Tuscany and Rome. However, Filiasi's affirmation of local identity need not have been hostile to national unification, especially, as was the case for Venetia, when joining Italy was tied to the need to break free of foreign domination. Indeed, since the fall of its republican government to Napoleon, Venice was first under French rule (1797–1815), then, except for a short period of renewed independence in 1848–1849, it was under Austrian occupation from 1815 to 1866, when it was joined to Italy by plebiscite.

Segments of the history of Burano can also be recovered from a number of reports ordered by different governments. Of these, possibly the most detailed was written in preparation for the Austrian Cadastre of 1827. It covered the areas of Cavallino, Sant'Erasmo, Torcello, Mazzorbo and Vignole, of which Burano was at that time the administrative centre. Even allowing that the hardships of the area may have been overstated to evade the Austrians' punitive taxation, the picture we derive is one of a truly impoverished and desolate lagoon. Because the inquiry was based on pre-established categories related to the prices of agricultural lands, rents and labour, passages on Burano, where at this period hardly any vegetable gardens or fields were left, alternate with paragraphs on its neighbouring islands. The nearby fish farms, which were Burano's main source of income, as well as the climate and the cycle of fishing, unfamiliar to Austrian officials, are described in some detail.

> The soil of the *valli* is a soft mud...in which eels multiply, without any need of work, and in which even white fish can find adequate nourishment, but, although the climate is usually temperate, when the northern winds clash violently against the sweet and salty Sirocco, they cause severe storms which may break down the reed walls and embankments of the fish farms, and cause great losses.

To make a precise estimate of the size and productiveness of the fish farms was indeed very difficult and the only certain information was that their rent was very low.[21]

> Those with a sandy bottom, shallow water mirrors and few and narrow canals...have very little wild fowl and in them the new fish has little success because they are a considerable distance away from fresh waters, while in the autumnal season those exposed to Sirocco are liable to have their locks torn apart by the winds and be thrown open, so that all the eels and the fish are lost.

During the long period of Austrian occupation, which ended in 1866 when the Veneto was joined to Italy, and the Italian Wars of Independence, the whole of the Veneto was dramatically impoverished. In particular, Venice, especially after the 1848 uprising, suffered outbreaks of cholera, while its peripheral islands became increasingly cut off and desolate. For Burano the nineteenth century was a time of increasing isolation and, despite the revival of lacemaking in the latter part of the century, its economy only began to improve after World War 2. Before I come to that, however, I shall return briefly to their history and to the debates surrounding its annexation to the Venice commune as they emerged from government reports which illustrate a vision of Burano from the centre at a time of political change and authoritarian centralisation.

From Foreign Occupation to Internal Colonialism.
Three Government Reports

Soon after Venice and the Veneto joined Italy, a need was felt to redraw their internal boundaries that Napoleon's classification of departments, municipalities and communes had rendered rather vague and ill-defined. The inland cities, which had formerly been under Venetian rule, claimed their surrounding territories and demanded a more equal position with their erstwhile capital than they had held before 1797. Venice thus seemed to have contracted and to have remained isolated within the natural confines of the lagoon. Moreover, a need to reorganise its internal administrative structure led to a drawn-out and sometimes acrimonious debate over the opportunity of aggregating or else excluding its peripheral islands. The background was Article 74 of the 1848 Statute of the Kingdom of Sardinia, which stated in a very broad and general way that 'Communal and Provincial Institutions are regulated by the law', followed in 1865 by a law which granted that it was allowed for a commune to *spontaneously* join another.

Legislation on the running of communes, which fully reflects the difficulties posed by a direct confrontation between the central state and local entities, was in fact characterised by highly contradictory tendencies. On the one hand, attempts were made to grant some degree of autonomy to local governments by extending the administrative vote to larger categories of people and establishing that mayors were to be elected. On the other hand, the laws were clearly formulated in the context of postunitary tendencies towards rigid centralisation.[22] Statements that fractions or communes could *voluntarily* fuse to form larger units, or else separate *by Royal decree* thus appear inconsistent.[23]

To further illustrate Burano's relations first with Venice and then with the newly formed Italian state, we must go back to the postunitary period, when it was first proposed that the island should be joined to the city's municipality – a proposal rejected by the Venetian council (24 July 1878) on the grounds that such a union would only have been a burden. The first relevant document is a report by a lawyer, Marco Diena, to Venice's Provincial Government, which was to mediate between the local authority and the central state represented by the *prefetto*. Its exposition of the issues involved and of arguments made in favour or against aggregation of different islands illustrates the tensions at the heart of relations between Venice and its periphery – some of which are still echoed in today's political and administrative interaction.

As emerges from Diena's report, not all islands were prepared to forgo their communal independence, and not all were equally acceptable to the Venetian administration. On the one hand, Malamocco and Murano had firmly refused to become part of the Venice commune (29 December 1877) despite the view of the city's municipal junta that 'thanks to a convergence of material, civic, and economic interests, it appeared entirely proper that they should be joined to the city.' On the other hand, when, in 1878, Burano's representatives applied to be joined to the Venetian municipality, their proposal was rejected with twenty-two votes against and only thirteen in favour (Diena, 1880).

To avoid appearing contradictory, Diena first established that each of the three islands gave rise to different considerations. Much of the discussion revolved on the definition of a 'walled commune', since that was the category named in the law as the most likely to benefit from aggregation of its outlying fractions. Could Venice be considered a walled city, as rhetorical comparisons of its canals with defensive walls had often implied? The main separatist argument put forward by the councils of Malamocco and Murano in defense of their autonomy was that Venice could in no way be considered a walled city. But, as Diena argued with a painstaking pedantry that almost looks like self-parody, the law could not cover a topographical condition as singular and out of the ordinary as was that of Venice: its case was neither excluded, nor touched on, nor was it even contemplated in the terms of that law.

> Venice is neither, in a literal sense, a walled commune, nor is it in the same condition as a commune that is not walled, but is like a village or city with a surrounding territory...No fiction or metaphorical language can have the powers to change the waters which surround our built areas into a true external territory, and, although Venice does not literally find itself in the condition of a castle surrounded by a wall, it is, beyond a doubt, in a completely analogous and equivalent condition, and the equivalence and analogy, in view of the aim and the spirit of the law, can in this case be invoked quite plausibly (11–12).

Because of Venice's unusual environment, therefore, the law was entirely open to the discretion of its interpreters. The city's refusal of Burano's request to be joined was sufficient to make its aggregation legally impossible. The Venice commune could not be forced to accept Burano's proposal; its application was declared legally inoperative and the refusal was confirmed by the Provincial council.

Concerning Murano, it was Diena's conclusion that the determination of its inhabitants to remain independent was to be respected. 'The subject,' he writes, 'was examined from a civil and moral, even more than a material, point of view'.

However, his arguments in favour of aggregating Murano (which eventually prevailed in the 1920s), were entirely practical and economic ones. In the first place, Murano was very close to Venice, one of the city's ports, Sant' Erasmo, was in its circumscription,[24] and, above all, Murano's numerous and productive glass factories were largely supported with Venetian capital (ibid.: 16–17). On the other hand, no reason was strong enough to take away Murano's autonomy, and a forced aggregation would have implied 'a complete abandonment of Venetian tradition' (ibid.: 28).

As for Malamocco's refusal to be joined to Venice, Diena's conclusion was that in that case the reasons for aggregation were quite overwhelming. Indeed, Malamocco's comune included the Lido, with some of Europe's best beaches, where private Venetian investors had recently built modern bathing resorts. Malamocco also gained importance and prestige from its port, with customs, telegraph, and civil engineering offices, all of which were supported with Venetian capital. What is more, its territory, a long strip of beach, with its dams and its two port openings, was like 'a seaward circle' and 'a natural bulwark round the city'. The Venice commune already owned the Lido's main street, which led directly from the lagoon to the sea front, and large sums of money had been spent on its lighting, tending its flower beds, building its bathing establishments and, not least, on policing; order and hygiene demanded that the territory should come fully under its aegis. For all those reasons, Malamocco was perfectly suited to the terms of the law; on joining Venice its land would at once rise in value, and the populations would certainly benefit and enjoy better hygienic conditions.

Here then, the reasons for or against joining the islands to Venice are clearly in evidence. With the exception of Burano, where a spirit of independence was eroded by poverty, inhabitants of the periphery at first were determined to defend their autonomy. In the case of Murano, awareness of being the main centre of the glass industry gave rise to the strongest opposition to the idea of being absorbed into the city's municipality. On the part of Venetians, however, the main reason in favour of Murano's aggregation was clearly the desire to protect their investments in the glass industry by administrative and political control. As compensation for their loss of autonomy, Venice offered the Muranesi the benefits of modern health services and infrastructures, like electricity, fresh water and so forth; a pattern of exchange that was repeatedly and persistently proposed, util the completion of communal unity in 1926.

Arguments in favour of aggregating Murano and Malamocco were evidently based on economic expediency, as were those alleged to justify the rejection of Burano; the island offered no hope of gain or profitable investment: it had no land, it was not a port, and, more important, it held no promise of economic development. As one of Venice's past mayors observed: 'Burano's real estate yields nothing; the landowners do not pay their taxes, and properties fall into the hands of the revenue, which has no use whatsoever for them' (Barizza 1987: 122). Its distance from the centre may also have been a relevant factor. In other words, Venice, at that time, simply had no use for that remote and impoverished island.

The proposal that Burano and Murano should be joined to the Venetian municipality was taken up again in the 1920s. A policy of internal colonialism, in keeping with an increasingly autocratic political climate, would both have favoured the realisation of an industrially developed 'Greater Venice' and allowed full control over the territory, so that no opposition might develop and simmer in peripheral areas, especially among Murano's socialist glass workers.

Before carrying out the desired annexation, inquiries were conducted to sound out the feelings and opinions of the people concerned and acquire direct knowledge of their living conditions and economic potentialities.

The two reports I am about to summarise provide a telling detail in the history of the rise of Venetian fascism. The first report was the result of a preliminary inquiry in Murano and Burano by a lawyer, Chiancone, in 1924. The second report (also 1924) was by his superior, Davide Giordano, a surgeon who was elected to Venice's council in the administrative elections of 1920, and subsequently called to manage its 'extraordinary administration' while the commune was temporarily commissioned between April 1923 and July 1924. Like the Austrian government inquiry of 1826, Chiancone's report includes the whole of the area from San Giacomo della Palude to the mouth of the Old Piave, of which Burano was then the administrative centre.[25] Rather untypically, Chiancone describes Burano as not too badly off. As he writes:

> The population of the chief town is mainly composed of fishermen, who in the past led a miserable life, while today their conditions are improved. The products of fishing, drawn especially from very extended *valli*, are sold in Venice. But the greater well-being of the village, is also due to the development of the industry of needle lace, which is very well known throughout the world, and in which are occupied about 800 women, who earn a decent daily wage. (15)

In the Burano commune there were altogether five parishes and three graveyards. Seventeen elementary schools were scattered through different hamlets. The school buildings were in good condition; their state of hygiene, as well as their balance of accounts, were reasonably sound, and pupils registered as poor were given free books and copybooks. Burano also had a nursery school, in which were gathered about two hundred children, both boys and girls. The only charitable institution was that of the *Congregazione di Carità* which had no income of its own and received a yearly subsidy from the commune. In a room which was municipal property, and under the guidance of the parish priest, there was an old peoples' hospice, or rather a dormitory, where about fifteen homeless persons were sheltered at night.

Burano only had one main road, that which joined it to Mazzorbo, but its state 'was no different from that described in the Cadastre of 1826'. The provision of water was still inadequate. Although two deep artesian wells dug in 1913 gave chemically pure water, they were not really sufficient for the needs of the population. Burano did have public lighting. The energy was provided by the Italian Society for the Utilisation of Hydraulic Forces in the Veneto, for an annual payment of 6,200 liras, but, as Chiancone writes, in a way that reads almost like a threat, that contract was going to end on 31 December 1923, and, had they refused to be joined to Venice, Buranelli might literally have been left in the dark. Moreover, they might soon have had to pay heavy taxes on staple foods, since the city intended to increase taxation by applying duty to a large number of goods, such as oil, sugar and coffee, which so far had been exempt.

Touching on an important area of fascist concern, that of the bureaucracy, whose members were often instrumental in exercising control over peripheral areas because of their eagerness to gain access to power structures at the centre, he describes the personnel of Burano's commune, two secretaries, an archivist, two clerks and a typist, in addition to its sanitary staff, which consisted of three doctors, a veterinary, four midwifes and a health supervisor. To ensure their loyalty and support it was emphasised that, being in the employment of the state, they were all entitled to receive pensions, while their 'subalterns', a messenger, an urban guard, three burial officials and four dustmen, could benefit from state insurance.

Burano's finances, while 'not of the most flourishing', were not such as to cause any serious worry. As in many areas that had to be evacuated during the First World War, government subsidies had been stopped in 1919 and expenses had to be reduced; some communal employees had been laid off, and local taxes increased. Indeed, as Chiancone writes, contradicting his earlier statement that Burano's finances were reasonably sound, the situation was altogether precarious and made life very difficult for the administration. More positively, however, he predicted that, thanks to fascist plans for the large-scale reclamation of marshes, the commune would greatly benefit by the drying up of a vast coastal zone from Tre Porti to the mouths of the Piave, which, 'would be transformed from fruitless swamps into fertile fields and fish farms'. A new land register would be compiled, and consequently the value of land would soon be more than quadrupled. A general revision of real estate would also bring in more taxes, and, with the rent increases that had taken place in previous years, the total income for Burano's building patrimony would certainly be trebled. The commune would then be in a position, 'to give to public services all those impulses towards renewal suggested by reasons of hygiene and public utility'.

The reason for the writer's eagerness to bring about Burano's annexation was mainly the fact that, since Murano was located on the route from that island to Venice, its joining would have made the case for annexing Murano all the stronger. As he writes, drawing on a repertoire of phrases and references dear to fascist rhetoric, when, fifty years earlier, Burano had asked to join 'its mother city'.

Venice had not understood the advantages of its domination over the lagoon. But the lagoon itself now feels the need of a powerful hand to protect its well-being and watch over the interests of the city and the port. Today, when Venice can better consider the benefits of expansion, it certainly will not want to refuse to grant its sister island, Burano, the embrace it is asking for…With Torcello, Mazzorbo, and San Francesco del Deserto, Burano is the best gem in Venice's crown. (18)

According to the report, the decision of Burano to join Venice (19 September 1923) was reached unanimously. But, as I was told by some Buranelli, who in the 1980s were still deeply upset at the memory, in reality the population had been uncertain and divided. The strongest opposition had come from the Popular Party, but, when its representatives saw that their views were ignored or forcibly suppressed, they walked out from the council room in indignation, so a proper debate never took place and the alleged 'unanimity' was reached at the expense of a correct democratic procedure. According to Chiancone's report, asked to explain the reasons why they were against annexation, 'they were not able to express their opinions'. A few days later they sent a written memorial,[26] so that nothing remains to be evaluated or refuted'. In Chiancone's view, therefore, resistance to the proposal of joining the Venice commune, 'found no echo in any person of the main town of Burano, and even less in its hamlets – a truth also confirmed by the parish priest, who pointed out that by their objections, the councillors of the Popular Party were not obeying any aspiration, other than that of continuing to occupy their seats in the council' (18).[27]

Ending his Report on a more positive note, Chiancone states that the main requests of the population were that some administrative functions, such as the registration of births, marriages and deaths, should continue to be carried out in Burano so that people should not be forced to travel to Venice for those acts, which they were used to carrying out locally. Also they asked that the personnel of Burano's commune should be allowed to continue in their jobs with no reduction of salaries. Other requests were for transport to be improved by the addition of two steamboats and the return to the timetables in operation before the First World War, which included ten daily return journeys. They also required that the construction of six artesian wells to provide drinking water should be completed, and that the list of the poor of the commune of Burano should be compiled with the same criteria as those adopted for Venice.

Touching on a more personal note, and disclosing patronage links on which hinged political power and influence, as well as a convergence of interests that could effectively join periphery and centre, Chiancone asked that persons with whom he had come into contact during his investigation should be permitted to keep their jobs; in particular, he asked that 'the secretary – who had declared that he would have been very happy to serve – should be continued as head of Burano's communal offices' (20).

While Chiancone's Report on Burano was based on the careful collection of data, predictably, his conclusion on the views of inhabitants on annexation of their commune to Venice was politically motivated. The same is even more evident in the pronouncements of his superior, Giordano. The factual basis of the two Reports is largely the same, as are their ideological premises. Giordano's Report, however, concerns the whole of Venice, and, with its aggressive patriotism, wordy rhetoric and unlimited use of classical references, it provides a vivid illustration not only of relations between Venice and Burano, but of those of Venice with Italy during the period of transition between Mussolini's March on Rome (October 1922) and the firm establishment of fascist control over Italian cities and municipalities.

Recalling that he had been appointed to take charge of the commune at a time of political disorder, 'a most evil assault on the Italian motherland', Venice, he writes:

> was one of the first great Italian cities to rise up against such danger. The parties of order and patriotism then joined together to form a bloc, of which the core were the National Alliance and the Fascio. A small group of brave young fascists, careless of danger, then, won that beautiful battle... Some who were not Venetian shed their blood here, while Venetians shed theirs in the streets and piazzas of other cities (3).

Here then, upholding the value of national unity, Giordano condemns both Italian localism and socialist internationalism, adding a strong note of contempt against those who 'wasted their time in empty polemics'. When he was called to run the commune, the main legacy left by the previous junta was one of financial and administrative disorder, 'while the dying counter-currents of the demagogic storms which had raged with deplorable violence' were making all council meetings convoluted and ineffectual. [28]

A strong theme in Giordano's writings is that of the Venetian state as precursor of Italian imperialism; its symbols of conquest and 'sweet domination' had been revived as 'the Winged Lion again roared in the square of Gradisca, beside the Roman Wolf'... and the long-lost amity between Venice and the cities which were once its subjects, Gorizia, Pola, Cormons, Fiume, Rovereto, Zara, was restored, as 'they all remembered with passion and nostalgia its ancient, glorious and just domination.' [29]

Concerning the estuary islands – in this context Pellestrina – Giordano again criticizes earlier administrations for not accepting 'with immediate spontaneity' its request 'to return to the breast of its ancient mother' (28 March 1920). Hesitation, he explains, was due to a need to weigh the 'financial passivities' which the island would have added to the balance of the commune, but not acceding to Pellestrina's 'legitimate' desire would have driven it into the sphere of Chioggia, whose periphery would then have extended in the direction of Venice as far as Alberoni. It was that argument, he adds, that induced the Venice council to 'magnanimously approve the aggregation of Pellestrina, which will open the way for further expansion of the commune, and restore to the city its ancient domains in the lagoon'.

When he mentioned this in the council, his speech was greeted with incredulous smiles, but the opportunity to carry out his plans had come much sooner than his opposers had anticipated, since it was favoured by the

Government's resolve 'to infuse into the vital organisms of large municipalities the rudiments and larvae of neighbouring communes struggling in the miseries of their penurious lives or torn by conflict and petty personal ambitions'. When the mayors of Burano and Murano were asked by the *prefetto* what their views were about the fusion of their communes with Venice, both said they were personally favourable, but the mayor of Murano feared that many Muranesi would be against the union – according to Giordano, such reluctance was due only to political instability (ibid. 12–13).[30]

As for Burano, Giordano naturally realised that the financial balance of Venice's commune was not going to gain from its annexation, and the island would be a burden and a source of expense for many years to come, but the benefits would be long-term ones; 'Those who were protesting would pass away, while Venice would remain in all its greatness'. It certainly was the city's purpose to remain prosperous and healthy, but it also was its mission to 'morally redeem ... those miserable areas undermined with malaria, poor in water, and only rich with wine cellars'.

Works had already been initiated to bring electrical energy to 'the trusting and industrious population of Pellestrina'.[31] A plan to provide drinking water to Burano, Vignole and Sant'Erasmo, whose inhabitants had complained about delays, was going to be examined. Naturally, huge sums would have to be spent. By acquiring new territory Venice would soon find new sources of income, but first the value of land needed to be improved, for, although destined for a prosperous future, large tracts were still unhealthy swamps. For example, in the recently annexed area of Mestre, large expenses would have to be met to reclaim land that was infested with malaria, but 'the money sown in the lagoon, and its desolate littoral, one day would flower [sic] with vegetable gardens, and villas would be built at the side of a new road, all the way from Punta Sabbioni to the mouths of the Piave'. Fortunately, to sustain Venice's economic renaissance, Mussolini and his finance minister De Stefani had canceled its war debt (eighty million liras), but careful planning and considerable investments were still required to give life to 'a greater Venice'.

Looking at the commune's plans for works, and examining lists of contracts for the years 1923 and 1924, one is struck by the fact that, although the measures proposed appear to stem from a need to exercise total control and domination over peripheral areas, priorities show sound business sense and a clear tendency to favour the more promising enterprises; the largest sums were in fact invested in Marghera, and the second main focus of expenditure was the Lido, in particular the construction of beach huts, roads, bridges, and sewers. A relatively large-scale project for Venice was the partial filling in of the Sacca di Sant' Elena, where there was built housing for a new contingent of bureaucrats and communal employees, as well as a new *Piazza d'Armi* for the training of cadets, and for such sporting and military activities as were required for a sound fascist youth. However, there is no mention of any money spent in Burano, or in the northern Lagoon.[32]

In the following paragraphs, Giordano returns to the topics of hygiene and good order. He deprecates behaviour that clearly went counter to his notion of discipline, such as the refusal of Venice's musical band to perform on some state occasion, or that of 'selfish and greedy' shopkeepers, who went on strike because state imposed prices were making their economic survival difficult. In his enumeration of social evils, which the new order was proposing to cure, Giordano points out that consumption of wine was excessive, and the police needed reorganising, as did all public exhibitions and festas.

Some of the measures suggested by Giordano (who seemed to have turned his medical training to the most intransigent and self-righteous authoritarianism), applied equally to Venice and its lagunar periphery. Much stricter discipline was to be imposed on gondoliers, 'a notoriously troublesome class'. There was an urgent need to check the numbers of street vendors: selling of fruit, vegetables and eggs, which was a convenience to the population, could be allowed, but sellers of cloth, saucepans, etc., 'a class often entered by the least commendable elements', had to be treated with utmost severity. [33]

Applications for licenses would be controlled by functionaries of the *Ufficio d'Igiene*. [34] They would check if the applicant was domiciled in Venice, or if he had gone there to ply his trade, what his conduct was, from which factories he obtained his goods and where he stored them during the night. They would also examine the health conditions of his house, as well as his own and his family's. Some foodstuffs, for example, milk, would have to be removed from street vending, because it was impossible to watch over the sellers and punish them for its adulteration in Venice's narrow streets.

Restrictions in the issuing of licenses for street vending, based on information taken by local officials, were, of course, extended to the islands as well. Indeed, in Giordano's view of the new order, the office for Public Hygiene had a very prominent place. Although supervision was rendered difficult by the distance and the slow pace of transport, great attention would also be given to school-buildings, teaching materials and, especially in poorer areas, to the cleanliness of pupils.

The most formidable task of the *Ufficio d'Igiene* would be the fight against tuberculosis and alcoholism. Its efficiency would be put to a hard test in Pellestrina, Murano and Burano. There, as Giordano writes, 'we have to fight against endemic problems: hygienic-sanitary provisions suffer from grave lacunae, which it will be both a duty and an honour for Venice to fill.'

To make that work easier, he had ordered that some small 'disinfestation units', already in existence in the islands but left in complete abandonment, should be restored, a prophylactic service for infectious diseases should be organised, and a speedy means of transport should be purchased, to make the admission of patients in the city's hospitals easier (ibid.:157).

Many of the measures proposed were, of course, much needed and salutary ones, but, as my informants remembered, they were applied with such heavy-handed and patronising insolence that their mere memory is tinged with feelings of humiliation and anger.

Notes

1. Doubts and curiosity were also due to archaeological discoveries which challenged earlier datings of settlement in the northern lagoon.
2. Among the names in a list of Torcello's bishops in the *Chronicon Altinate*, are one Giovanni, the son of the tribune Aurio, and Stefano, both natives of Burano (Monticolo 1890).
3. In the past, it was sometimes said that Buranelli were 'orientals' because they supposedly had curly hair and sallow skin, or that they were the descendants of galley-slaves confined to Burano, and thought to be devious, thieving, litigious and particularly aggressive in canal traffic disputes.
4. References to Attila are also common in rural Friuli.
5. A visual analogue may be 1960s narrative 'pop art'.
6. Travel writers give currency to such stereotypes. For example, as J. Morris writes, 'a special race of men, too, has been evolved to live in this place: descended partly from the pre-Venetian fishing communities, and partly from Venetians who lingered in the wastes when the centre of national momentum moved to the Rialto. They are the fittest who have survived, for this has often been a sick lagoon...Like the rest of the fauna [*sic*], the people vary greatly from part to part, according to their way of life, their past, their degree of sophistication, their parochial environment. In-shore they are marsh people, who tend salt-pans [these actually went out of existence at the turn of the century], fish among grasses, and do some peripheral agriculture...Their dialect varies, from island to island. Their manners instantly reflect their background, harsh or gentle. They even look different, the men of Burano... tousled and knobbly, the men of Chioggia traditionally Giorgionesque' (1960: 258).

    By contrast Dorothy Menpes describes Buranelli as 'statuesque': 'One sees boatfuls of them returning from the sea; and lines of them towing heavy mud-filled barges on the way to Pordenone, all the men stepping in time. With their long cleanly-moulded limbs, they remind one of ancient Egyptian bronzes. The sculptor would find plenty of scope in Burano. The people, however, are of evil repute by heredity. They are the scapegoats of the lagoons. If anything goes wrong, the blame is always laid upon them. They work harder and receive less pay than the inhabitants of any other islands. (1904: 165).
7. Whether *Giudecca*, in Venetian *Judeca, Zuecca, Zudecca* or *Zudegà*, derived from the Latin *Judaica*, or from *iudicare*, to judge, is a question that has vexed historians since the seventeenth century. Derivation from *Judaica* would support a view that Jews resided there since the twelfth and thirteenth centuries. While historians have looked to Constantinople, Crete and Negroponte for comparison, the fact that nearby Burano also has a quarter named Giudecca has not led to much comment.
8. Lazzaretto Nuovo was originally inhabited by hermits. In 1468, it was equipped to store goods that were feared to carry infections. During the plague of 1756, those who had been in contact with the diseased were sent there while waiting to be admitted to the old Lazzaretto, or dismissed if they were found to be healthy. Later, the island was used for military installations, and it is now left in a state of abandonment.
9. In a letter to a patron he describes himself as 'fruitless Calmo, brought up on fishing boats, fed in the baskets, and taught to capture fish' (Rossi 1888: 211). One supposition is that, after an early childhood in one of the lagoon islands, he may have been sent to a seminary, but may have abandoned a clerical career in favour of the theatre. That would explain both his familiarity with fishermen's lives and his intimate knowledge of aulic poetry and manners, which enabled him so effectively to satirise them. Written at an important turning point in the linguistic history of Italy and Venice, when Italian was increasingly used in literary writings in place of both Latin and Venetian, Calmo's works are a vindication of the great expressive potential of the dialect, or of the numerous dialects then spoken in and around the city.

10. Calmos' characters eventually developed into the stock figures of the *Comedia dell'Arte*, everpresent in the theatrical as well as the carnival traditions, which in turn contributed to fixing them in the popular imagination as enduring stereotypes (Padoan 1982: 16).

11. Corner was a devout and dedicated Church historian, by contrast Coronelli was a worldly and well-traveled Jesuit geographer, who also drew maps of Oxford and Cambridge (1702) and accounts of all areas under Venetian influence in the Adriatic and Aegean.

12. For example, by 1432, Filippa Condulmero, Abbess of Santa Cristina and of St Mark of Ammiana, was forced to abandon her convent and join the community of St Antony, as 'she had remained quite alone, and no other person wanted to take the veil' because of the bad state of the building'(Coronelli 1696: 33–34).

13. Mazzorbo had four convents, and two parish churches, with curates elected by the local landowners, as was the custom in Venice, as well as five women hermits and only one male congregation, that of the Celestine fathers. Indeed, except for St. Francis of the Desert, in the island which bears that name, the islands seem to have been dominated by communities of religious women.

14. In San Vito there were twenty-six nuns, eighteen choristers, and some lay members, while in the 'Capuchins', founded in 1626 by pope Urban VIII, the total number was twenty-three, fourteen officiants and nine lay sisters.

15. That, Corner writes, was 'A most dismal time when the main powers of Europe [united in the League of Cambrai] were plotting against the Republic...and enemies were ferociously overrunning Venetian lands, filling every place with ruins and destruction. To preserve their honesty, which they valued even above their lives, from military insults, some afflicted nuns, women of the most upright and religious life, fled to Venice from the convent of Santa Maria della Misericordia in Noale, a castle in the territory of Treviso...In 1516, on the day of the martyrdom of Saint Vito and other saints, 15th of June, the nuns were granted the ancient Priory of the Saints Cornelius and Cyprian, which had been abandoned and gone into *commendam* since time immemorial' (1759: 600).

16. Anticipating criticism, Filiasi acknowledged that his work was speculative and his sources fragmentary. For example, he refers to a tradition derived from the *Iliad*, according to which, during the Trojan wars the Heneti would have fought against Athens. The reference is actually not an extensive one, 'Pylamenes of the shaggy breast led the Paphlagonians, from the lands of the Heneti, from which come the wild mules (trans. E.V. Rieu, II, 1977 [1950]).

17. Following Filiasi's view that 'language is the most indelible mark of all nations', Romanin also describes the lengthened vowels of Burano's dialect as 'a remnant of the pronunciation of the *Veneti Primi*'. Their early marriage customs are thought to derive from those of Babylonia, but mention of Oriental origins is qualified with remarks that, after the Veneti settled in Italy, 'under a different religious form, their customs had changed and they had abandoned their Asiatic laxity' ([1853] 1972: 30).

18. To support this, Romanin quotes a twelfth century song in Buranello dialect,

> 'Che me mario se n'é andao/ Chel me cor cum lui á portao/Et eo cum ti, me deo confortare.'
> (Because my husband has left / and with him he has taken my heart / with you, now, I must comfort myself) (1972:13).

19. Mention of aristocratic origins illustrates the difficulty of reconciling the islanders' poverty with assertions by early chroniclers that all the settlers were descendants of aristocratic Christian *Gentes* – a problem we often find in Venetian historiography, and which probably finds an answer in the hypothesis that the islands may have been inhabited before the arrival of refugees by *Veneti Primi*. In that view, Buranelli, like isolated communities of peasants or mountain dwellers, are ennobled by antiquity, as they are thought to embody some essential and uncorrupted primordial character lost to other Venetians. Despite being viewed as an exciting find by the linguist and ethnologist concerned with ethnic identities and origins, however, Buranelli have often been described as backward and uncouth.

20. Filiasi refers to Captain Cook, and in his analytical index he describes 'O-Taitians' as people who hold the belief that 'the dead intensely desire to drink the blood of their living relatives'.

21. Rents were no higher than six or lower than four Venetian lire for the poorest *valli*. Prices for basic staples were roughly the same as in Venice, but labour costs were slightly lower, and the wages of men were always higher than women's by amounts that vary between 15 percent and 80 percent. A man's average daily pay was 2.10 lire – it could sometimes be as high as four Venetian lire in summer and three in winter – and it included drink, salt and wood for heating.

22. One such law was that promulgated by Rattazzi in October 1859 (*La Costituzione Italiana,* Article 74. 1975: 56). Further legislation on local government was issued in 1862, 1865 and in 1888 (30 December, N.5865).

23. In some instances te redrawing of boundaries could be brought about by state authority, provided it was carried out within five years. That deadline was later postponed by five more years. As Ghisalberti writes in his Constitutional History 'After centuries of political particularism, and institutional pluralism, aggravated by foreign domination, as well as lay and ecclesiastical tyranny... this structure had to be recomposed on new bases with the total commitment of the state apparatus. Hence the veneration for the state, the cult of its very idea and trust in its laws and institutions on the part of the leading exponents of moderate liberalism' (1985: 105).

    In reality, government vigilance and control were tenaciously upheld, both to keep checks on separatist tendencies, considered anti-national and anti-unitary, and to maintain at the periphery the political hegemony of the liberal bourgeoisie which then dominated. Statism (of a kind) reached a peak with Mussolini's March on Rome and the victory of fascism, when locally elected mayors were substituted by *podestá* nominated by the central government, while *consulte* made up of members of fascist unions took the place of communal councils, so that any form of local government was eliminated. It was soon after the establishment of a fascist government that several estuary islands were annexed to Venice.

24. As Diena writes, since the hamlets of Sant'Erasmo and Vignole, like Murano itself, were plagued with infections, and their ditches and drains were causes of malarial fevers, they were threats to the well-being of the larger commune.

25. According to the official census of 1921 the Comune had 9,574 inhabitants, and its extension was 22km in a direct line. It was subdivided into two census areas, Burano and Treporti, and it included the fractions of Cavallino, Mesole, Liopiccolo, Torcello, S. Cristina, Montiron, Cura, Mazzorbo, San Francesco del Deserto, and Valli Dogado, Grassabo' and Ca' de Riva.

26. Chiancone claims to have included that document with his own transcript, but it was nowhere to be found in the files which I consulted.

27. Proposals concern issues like transport, medical services and needs that are still not adequately provided for by the municipality, and are the causes of much resentment against the city.

28. As some Venetians remember, the walk-out of opposition councillors was prompted by indignation and by a feeling that all was lost, as much as by a fear of Black Shirts' violence.

29. The Lion, as Giordano writes, also 'crossed the Ocean to climb on to the façade of a church consecrated to St Mark, in the Brazilian city called "New Venice" by migrants mindful of their glorious Mother and Queen'. During Giordano's three years in office, the Ministers Mussolini, De Stefani, and Ciano were granted honorary citizenships 'because they loved Venice'. Others were the historian Arrengo, the Duchi della Vittoria, General Diaz and Admiral Thaon of Revel.

30. Murano's commune was in a state of deficit, one which, as Giordano threateningly reiterates, was going to deteriorate very rapidly, because with new arrangements for the collection` of duties, it would soon receive even less revenue than it had up until then. He writes, 'Murano had four political parties for a population of barely 5,000...As soon as they heard the news of the annexation of their commune to Venice, its citizens were put into a state of alarm by some of their leaders, or by persons who pretended to be that, almost all of them members of the Popular Party'. The main issues were a dire lack of services, a 'need of everything', and a deficit which would have got worse as long as Murano remained a separate tax area.

31. In reality the Electrical Company, Sade, had been overproducing, and at this time was having some difficulty marketing an excess of energy!
32. As Giordano writes, 'The decision to build a new parade ground on the Sacca Sant'Elena, the cancellation of the war debt and the 'gift' of the Doges' Palace are such great concessions of the Government of his Excellence Mussolini to Venice, that they fill me with intense gratitude…for having had the good fortune of acting as intermediary in actions of such vital interest to the city.'
33. Before 1923, there were over one thousand street vendors, but over three hundred licenses were withdrawn in a short time.
34. The *Ufficio d'Igiene* was responsible for a very broad range of repressive measures. 'Hygiene' itself, had become a highly ambiguous notion and had acquired heavily moralistic and didactic connotations.

BIBLIOTECA DI
BURANO

CONSIGLIO DI QUARTIERE
DI BURANO

"BURANO E IL SUO INTORNO..."

..., puzzle di storia,...

Nell'ambito dell'iniziativa già avviata riguardante la costituzione di un archivio storico locale, la
Biblioteca di Burano promuove nel paese la raccolta
di materiale storico documentario riguardante l'ambien
te,i costumi,le usanze di burano e del suo intorno la-
gunare.

Si invita pertanto la popolazione del quartiere, le
associazioni culturali a collaborare fattivamente all'
iniziativa mettendo a disposizione il materiale in lo-
ro possesso: vecchie illustrazioni fotografiche (di fa
miglia,di matrimoni,ecc.),cartoline,manoscritti,datti-
loscritti,documenti sonori dialettali e qualsiasi altro
materiale riproducibile che testimoni i modi di vita ,
le usanze,le tradizioni,la vita politica,il mondo del
lavoro,le trasformazioni sociali ed economiche avvenu-
te nel nostro qurtiere.

Tutto il materiale verrà riprodotto e gli originali
restituiti, asicurando l'anonimato delle persone ripro
dotte nelle illustrazioni.

Il materiale così raccolto costituirà assieme ad al
tro materiale la base per una prossima mostra antologi
ca; lo stesso sarà collocato nell'archivio storico del
la Biblioteca e potrà essere liberamente consultabile.

RACCOLTA DEL MATERIALE
Il materiale verrà raccolto in Biblioteca nell'ora-
rio di apertura e dovrà essere consegnato in busta chiu
sa all'interno della quale dovrà essere specificato :
il Cognome, il Nome, l'Indirizzo e il Numero Telefonico
del proprietario dei documenti.
SI RINGRAZIANO FIN D'ORA, QUANTI VORRANNO COLLABORARE
ALL'INIZIATIVA.

Figure 2.1 The puzzle of history.

Figure 2.2 The arrival of Burano's Patron Saints, Albano, Orso and Domenico from the sea. Innocent children miraculously draw in the heavy sarcophagus. (Antonio Zanchi 1690. Parish Church of San Martino, Burano)

3
RELIGION AND SOCIAL CHANGE

For many Buranelli a sense of the past is imbued with religious memories and religious significance. Like other islands in the northern lagoon, Burano was settled before the Rialtine estuary achieved its dominance, and ancient chronicles, as well as Torcello's splendid monuments, are predominantly Byzantine and clearly recall early Christian, rather than Roman roots. In general, Venice's foundation myths imply that the populations who took refuge in the islands to escape from the repeated invasions of pagan barbarians did not do so only for the sake of their safety, but to be able to freely profess their religion and to safeguard their identity as Christians.

From Coronelli's and Corner's brief passages on Burano, Torcello and Mazzorbo, we form a picture of an area settled by Benedictine convents, and one in which numerous churchmen and clerics lived among the humble population of market gardeners, fishermen and lacemakers. And just as Venice was blessed with the translation of the spoils of the Evangelist St. Mark from Alexandria, also Burano had its miraculous arrival of three Saints' bodies, those of the martyrs Alban, Dominic and Orso. The saints' remains are now in the church of San Martino where the miracle is portrayed in a lovely canvas by Antonio Zanchi (1690), and the legend, which emphasises the purity and power of the young, was often relayed to me by Burano's schoolchildren.

Since the mid-seventeenth century, Burano also had numerous religious associations and brotherhoods, all of them suppressed in 1805 by Napoleonic decree, as were the Benedictine convents of San Vito and San Mauro, and that of the Capuchin Nuns of Santa Maria delle Grazie. Catholic associations were formally founded again after the First World War.

For the past, a picture somewhat contrary to that of a pious and totally Church-dominated community is conveyed by Calmo in his comical description and dramatic portrayals of the islanders. A small number of reports of Inquisition trials also present a contrasting tale of ribaldry and of radical divergence from Church-dictated conduct and belief. In more recent times too, as I found during fieldwork, defiance and rejection of ecclesiastical (as well as secular) authority were not rare. A young woman, for example, assured me that words such as 'penance', 'sacrifice', 'must' or 'needs-be' were not allowed to enter her vocabulary. A much older woman ridiculed the idea of virginal bridal veils for Mussolini's daughter, while a considerable fall in the production of lace (see chapter 7) was due not only to economic reasons but also to a general rejection of the view that women's working conditions should be based on concern with sexual purity and with a religious aesthetic that had come to be seen as empty rhetoric.

Referring to the etymological association of the Latin for village, *pagus*, with pagan, and indeed following the Church's survivalist views of local practices, one of Burano's priests complained to me that he was treated 'just as a ritual agent'. People, he said, certainly would not dispense with his officiating at baptisms, confirmations, weddings and, above all, funerals, but they did not listen to him, they were materialistic, miscreant and selfish. Indeed, as he said, implicitly referring to my anthropological background, he didn't have to go to faraway places to do missionary work, since he had to do it every day there and then in Burano! He, for example, found the Buranelli's celebrations on All Souls' day quite excessive and every year he had to remind them to attend mass also on the following day, the day of the saints; it was after all the saints, not the dead, who were the true community of Christ. When he officiated at baptisms, he had to really watch himself and spell out every word in a very loud and clear voice, because some people held on to an old belief that if the ritual was not perfectly performed, the child would grow up strange and even develop the rather dubious ability of seeing and communicating with the souls of the dead. However, he said, threats to his office greater than those of old superstitions really came from contemporary forces.

Such complaints were part of an ongoing dialogue between the priest and his parishioners, at a time of intense social change. Buranelli's distinction between 'tradition' or 'times past' and a rapidly changing contemporary reality was particularly in evidence in discourses about religion. Like other parts of Italy – and notably the Veneto, one of the most Catholic of Italian regions – throughout the 1970s Burano has undergone a degree of secularisation. However, despite the priest's occasional complaints about irregular church attendance, compared to Venice and Mestre, Burano in the 1980s was a relatively devout community.

A religious orientation was certainly present in children's socialisation – both in the family and in the earliest experiences of school, given that, as for most of Italy, the local infant school is run by nuns, while religious instruction and prayer (a much debated battleground between the Church and the secular state since Italian unification), although no longer compulsory since 1984, were usually present in the ordinary routines of Burano's primary and secondary schools in the 1980s. Sporting and recreational activities often take place on Church property, and games, especially for boys, are supervised by a young priest or by dedicated volunteers. Boys and girls are carefully instructed in Catholic doctrine before their first communion and confirmation, although gender differences become increasingly marked as people grow into adulthood.

As anthropologists have shown for other parts of Italy, women's devotion is strongly focused on the Virgin. In Burano her cult is rendered even stronger due to the association of the island's main craft, lacemaking, with chastity, patience, perfection and self-sacrifice, all virtues best exemplified in the life of Mary. As we shall see in my discussion of historical accounts of the craft, speculative attempts to identify its origins have led nineteenth-century scholars to seek references and anticipations of it in scripture, and in particular in apocryphal narratives of the life of Mary. Thus, given that for Burano's women, including those who never made lace, that craft was itself a token of the island's identity, such associations of its beginnings with scriptures is part of an enduring discourse in which the Virgin is held as a source of inspiration and a paramount example.

While the Virgin's attributes show that her virtues are infinitely various, she mainly represents motherhood, while at the same time emphasising its mystery and its potential for love, as well as fortitude at times of suffering, as paths to redemption. Images of the Madonna, and sometimes small domestic shrines in which her effigy is placed on a shelf or sideboard close to those of dead family members and lit by a tiny electrical bulb, are thought to infuse strength and courage in accepting the sorrows and disappointments that are thought to be almost inevitable in a woman's life. In a crisis women will always turn to the Virgin to ask her to intervene in their favour and their defence. At the time of my fieldwork, girls were encouraged to make a lace figure of the Virgin, or – easier to execute – a lily, itself a symbol of purity, in preparation for their first communion. Their handiwork would then be framed and examples are found in almost every home in Burano. As a devout woman explained to me, for her the Virgin had been a source of comfort and inspiration throughout the various stages of her life: as a symbol of chastity she had helped her to overcome temptation, when in her early youth she had been far too playful and flighty, she had sustained her at times of loneliness and discouragement when her fiancé was a prisoner of war in Africa. When she struggled to bring up her children, her image of the Virgin had changed from that of a loving mother to that of an exemplary sister and fellow-sufferer.

Shrines with effigies of the Virgins and Saints are spread throughout the island; of particular interest are those referred to as 'Dressed Madonnas', or, more precisely 'Madonnas *to be dressed*', of which Burano has two, the Virgin of the Rosary and the Virgin of the Snow. The first wears a white tunic with an elegant lace front, while the second wears a sumptuous white dress, it too decorated with fine needle lace (figure 3.2). Every year, before their feast day, the effigies are taken to the church, where devout virgins renew and refresh their clothes, behind strictly closed doors (cf Pagnozzato 2002: 35–40).

Several women told me they had particular faith in the *Madonna Bambina*, the Virgin of the Nativity, whose popular cult has long been widespread throughout the Veneto. For example, in the fishing village of Caorle, not far from Burano, the festa, on September 8, marks an important turning point in the yearly cycle and is solemnly celebrated every five years, when the Virgin's effigy is taken out from the Cathedral at the head of a procession that follows it through the streets. It is then taken out to sea, and accompanied by a fleet of fishing boats to the sound of songs in her praise.

The cult enjoyed a revival thanks to the foundation of the nuns' order of *Maria Bambina* at the turn of the eighteenth century, and its association with Burano was made stronger when nuns of that order were put in charge of the Lace School and of the nursery school. As one of the nuns explained to me, initially the order did not receive much encouragement from Rome, but eventually its founders, Capitanio and Gerosa were canonised as Saints. Partly the reason for the order's initial difficulties was the fact that its effigy was easily confused with images of the baby Jesus (figure 3.1). In its favour though, the nun continued, it is a very appealing image for the simple and the young, indeed, some of the more devout women keep a small replica of Baby Mary under a glass dome in their bedrooms, and they assured me that She had brought about some miraculous cures.

While God the Father is addressed in prayer every day, it is Jesus and Mary who are most often invoked and the sentiments described to me are usually modeled on family ones; men also worship the Virgin, whom they regard as the epitome of motherly love – a protective figure who reinforces their sense of purpose, helps them to achieve success in their enterprises and pleads for God's forgiveness on their behalf. Like all Venetians, Buranelli are devoted to the Madonna della Salute, who, in 1630, delivered the city from a devastating plague, and every year, on 21 November, the feast day which celebrates the Presentation of the Virgin to the Temple, many of them take their families to attend Mass in the church built in her honour. A fisherman who had become a champion rower in Venice's regattas told me he never forgot to cross himself when he rowed past that church, even if that momentarily slowed him down, because he was sure that the Virgin had supported him throughout his sporting career, and she would again help him to gather new vigour and overtake his rivals in the latter part of the race.

Women's devotion to Jesus was based on intense and lasting intimacy. Transposed on the plane of marital and family relations, this was sometimes viewed as problematic in Italian life, especially by psychoanalysts and social workers who read in men's identification of the sexually forbidden mother with the Virgin a configuration of oedipal sentiments, which might greatly prejudice the success of later relationships (Parsons 1967: 381–387, Goddard 1996: 189–204).

The devotion of men, and most especially fishermen, is also addressed to the island's Patron Saints, Albano, Domenico and Orso, while at moments of danger at sea, and especially in the 'old days', sailors and fishermen appealed to the souls in purgatory, who are thought to be suffering torments and anxieties comparable to theirs. Older Buranelli usually refer to the main turning points in the year's cycle and fishing calendar by the names of saints. For example, given that parts of the lagoon are known to be richer than others in new fish, and places for fishing used to be assigned by drawing lots on the day after St. Peter's day, in early March, in describing their good fortune or defending their rights against others' intrusion, men would say 'this place is mine: it was given to me by St. Peter'. Shooting quails in the marshes, they told me, was best 'between Madonnas', that is, between the Virgin of the Rosary on 15 August and the Nativity, on 8 September, the official opening day of the duck shooting season. Early September was also an important turning point in the fishing cycle, since on the day after the Nativity the fishermen, who often referred to the *festa* as the 'Madonna of baskets and bundles', *dei fagoti,* would move to their temporary shelters (*casoni*) for the Autumn fishing season, which lasted until 23 December, just before Christmas.

Church attendance in Burano, while not always coupled with strict doctrinal commitment and belief, is traditionally a basis for belonging. This was all the more so in the past, when, as my informants recalled, going to church on Sundays made a lovely change. After a week of toil, it gave people a reason to wear their better clothes, to walk across the main street and piazza and exchange greetings with neighbours and friends. For many women the church was the only place where they could go to see and be seen by others and it fostered a strong sense of community. However, as I found in the course of my fieldwork, while many peoples' lives and discourses were firmly rooted in religious beliefs, principles and customs, in the 1980s atmosphere of questioning and social change some of Burano's younger generations were naturally drawn towards secularising tendencies widespread throughout the country since the 1960s and 1970s.

Regular church attendance was no longer considered an essential basis for reputation and morality, while secular values and experiences competed with religion in guiding actions and influencing decisions. Lifestyles, and in particular political choices imposed by the Church, were often contested or reconciled with other models. An argument that the ideals of social and economic justice were rooted in Christian doctrine long before they had been upheld by left-wing political thinkers (Hirschon 1989: 193) was sometimes refuted, especially by younger Buranelli, who said that an attentive scrutiny of contemporary realities and politicians' actions clearly showed it to be facile and untenable. In particular, the hegemony of the Christian Democratic party, which was largely due to the support of the Church, was beginning to be strongly questioned, although, before the 1990s '*tangentopoli*' financial scandals,[1] many people in Burano thought it best to belong – and be seen to belong – to that party (cf. Pardo 1996: 31).

This was clearly brought home to me when, on one 25 April, Burano's streets were enlivened by different and overlapping celebrations: it was at the same time the feast of Venice's patron, St. Mark, and the anniversary of the Liberation at the end of the Second World War. The two celebrations, the *Liberazione*, which is usually associated with left-wing forces, but is nonetheless valued by most people in as far as it was a rejection of German invaders, and San Marco, as a strongly local *festa* with both religious and secular overtones, are not mutually exclusive and antagonistic, but they take place side by side. Political speakers often underline the coincidence of the Liberation with St. Mark's feast as a happy one, and both contribute to a festive atmosphere and the celebration of Spring.

On the day of St. Mark it is the custom in Venice for women to be offered a rosebud by their closest male relative, while carnations are handed to passersby by socialist party activists. Not fully aware of people's sensitivity to the power of symbols, I unwittingly provoked a revealing contrast and a momentary irritation between a usually devoted husband and wife. That year, while flower vendors were busy selling carefully cellophane-wrapped rosebuds in the Piazza, a small group of men who had attended a socialist meeting were handing out free bright red carnations to all passersby. I too was given one and, as I was walking back towards the boat, I was joined by a friend, Gino, who persuaded me to join his family for a drink. He was holding several roses and he at once offered me one, while I, already burdened with my briefcase and camera, asked him to carry both the rose and my carnation for me. As we arrived at the house, his wife reprimanded him, 'what are you doing, walking about with *that thing* in your hand?' I must have looked puzzled, while, taking the roses from him, she continued, 'these are what I like'. Then, as Gino stood smiling in embarrassment with the carnation still in his hand, she conceded 'that too, poor thing, is a flower', and, after cutting its long stem, she put it all alone in a glass vase, well away from the roses, 'it is quite nice really, a nice flower, but I'll have nothing to do with those people!'

An undisputed allegiance to the Christian Democrat party and regular church attendance, however, were increasingly questioned, while fear of others' judgment was by no means as strong as in the past. Beliefs and morals were considered independent of a total acceptance of Church dictates, but such liberalising tendencies were often the cause of intergenerational conflicts and antagonism. For example, for Teresa, who was determined to see Church and Christian Democracy as inseparable, her son's fascination with left-wing ideology was a constant source of family tension and debate. While frequenting a Venetian lycee the boy had formed a close friendship with Franco, a youngster who, from Teresa's point of view, summed up all that she regarded as dangerous and unsettling to her deeply-held values: Franco's parents, both members of the Communist Party, were separated and planning to get a divorce. She thought that they were most probably given to drink (as, she suspected, were most communists) and she feared that his exalted political ideas and speeches may have been enhanced by drugs. When her son, Marino, one day came home from school with a political poster marked with a broad red line over its heading, she not only reproached him for bringing such *trash* to the house, but ordered him to hide it immediately, especially as it was near Easter time, and the priest was due to come and bless the home.

The main area of dissidence from Church teaching, also revealing of changing attitudes to honour and gender, was that of sexual behaviour. An old saying that 'each person is master of his/her own body' was increasingly quoted in support of widespread liberalising tendencies. Several women assured me that, contrary to the Pope's views, and to the disapproval of the parish priest, contraception was 'the best thing that had ever been invented' and very few agreed with the Church's attitudes to divorce and abortion, although they would not have wished either to have occurred in their own families, except in extreme circumstances. Since secular experiences and values now competed with religion in guiding actions and influencing decisions, there was a feeling that marriage and the family should be secularised and young couples, as well as individuals, allowed much greater autonomy than in the past. In general, there was a tendency to withdraw from regular and obligatory church attendance and to view religious experience as a private matter.

Attitudes, ideas and beliefs absorbed through centuries of Catholic teaching are nonetheless an important part of many Buranelli's consciousness. Indeed religion is difficult to separate from other aspects of life, especially kinship, honour and, for Burano, the history of lace craft. My account and discussion of it, therefore, will continue in later chapters of this book.

Note

1. Bribery scandals that discredited the Christian Democrats, as well as the Socialists and smaller parties.

Figure 3.1 Maria Bambina. 'Hear my prayer, oh heavenly girl, and I shall forever praise the goodness of your heart' (Reciting the full prayer will grant 100 days' indulgence. Cardinal D. Agostini 1885)

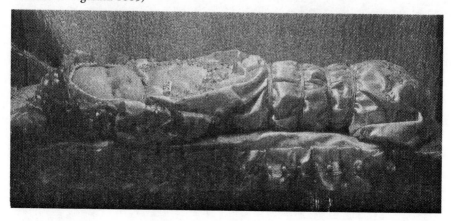

Figure 3.2 Madonna della Neve.

4
KINSHIP AND RESIDENCE

The relevant context of specific cultural elements, such as 'marriage', 'mother', 'blood', or 'semen', is not limited to current practices and meanings, but includes past practices and their symbolic meanings (Yanagisako & Collier, *Gender and Kinship*).

One of the social insights most commonly offered by my informants was 'here in Burano we are all one big family' – a fact which, they thought, would provide an essential basis and starting point for my study. Similarly, when asked by an English writer to describe differences between Torcello and Burano, a Torcellan woman answered:

> We are very different from the people of Burano even if we are only five or six minutes away in a rowing boat. For instance in Burano they say they are related to I don't know how many degrees of kinship. Those people set great store by the family. It means a lot to them and they hang on to them [*sic*] even when they are right outside all the recognised degrees. It must be because they have all been there since the world began…We, here in Torcello, certainly we love our families, our parents and children, cousins and aunts and so on, but family relationships matter less to us outside these degrees, but the Buranello really clings to them all. Everybody is related to everybody else over there and they like it that way (Guiton 1977: 163).

The woman's reply actually contains several statements: as well as a purely factual observation – Buranelli 'say they are related…' even to distant kin – it also conveys her sense of puzzlement and a hint of criticism: 'they set great store by the family', 'they cling' and 'hang on' to their relations and 'they like it that way'. Finally, her mention of Buranelli's attachment even to relatives 'outside the recognised degrees' inevitably carries echoes of reckoning and of tensions over marital choices. All aspects of kinship in the woman's account deserve a closer analysis. To begin, I shall describe residence patterns and the ways in which

Buranelli interact with and express sentiments about their kin, as well as changing attitudes and expectations. In the second part of the chapter, I shall examine Burano's kinship terminology, discuss aspects of socialisation, gender differences and the use of nicknames; to end, I shall show ways in which kinship connections and networks link people resident in different islands.

Residence

An old saying, 'bread, maize, and your own front door' (*pan, polenta e porta sola*) states Venetian preferences in the matter of household forms with cryptic finality. Indeed, according to the popular saying, the mere sharing of an entrance hall, staircase and lobby can be a source of irritation and friction. Should a new couple be in a position to choose between a meagre diet of '*pan, polenta*' in an independent home and richer fare in shared accommodation, they would not hesitate to opt for the former and sacrifice all unnecessary luxury in order to gain complete family privacy (Hirschon and Gold 1982: 66, 70).

A desire to start married life in a new home of their own, which most people now regard as an indispensable step towards independence and adult status, is certainly long-standing in Venetian tradition, but, while in the past couples might establish their nuclear households several years after marriage and the birth of children, since the 1960s and 1970s, young people have become firmly convinced that having a separate home is a right and not just a vague and remote aspiration. Therefore, concern about house prices and rent laws, which at the time of my fieldwork seemed to dominate Burano's life and political discourses, was due both to scarcity and to greatly raised expectations. As we have seen in my introductory chapter, young couples' desire for separate marital homes was strongly supported by modernising forces, in particular a tendency to put a curb on parental authority, to lessen the influence of the Church[1] and in general to support individual needs and rights.

A contrast between past residence patterns and present-day perception of needs appears very clearly if we follow the housing histories described by some of my informants. Nina, an elderly widow who always welcomed me on my afternoon visits, lived alone in a pleasant house near the piazza. She had lived there for most of her adult life, indeed, so great was her attachment to the house that she 'almost never left its four walls'. But that was not the way she felt in the early years of her marriage when she had often wished she could 'fly away, like the tiny sparrows that sometimes stood for a brief moment on [her] windowsill'.

When she got engaged in the late 1920s, she agreed that on marriage she would go to live with her husband's paternal family in their spacious house, with a little backyard, where she grew a few herbs and a lovely variety of carnations, geraniums and roses. The different rooms in the house were variously allocated and newly furnished according to changes in the family's composition and structure. A few days before Nina's wedding, her widowed mother-in-law moved out of the main bedroom, to honour Nina and her husband, who was the oldest of three sons. Nina then moved in with her specially crafted bedroom furniture, her loom and her wedding box full of embroidered linen. At that time the house was filled to capacity; the kitchen on the ground floor was light and spacious, as was the parlour, which in those days was often open to visitors. On each of the two floors above were two bedrooms, occupied by various members of the extended family.

In time, Nina's husband's brother with his wife and two children moved out to establish their own nuclear household. The second brother left when he married, so she was left in full charge of the house and of the care of her husband's mother, who had become an invalid – a very unhappy time, given the old woman's difficult and demanding character. The household, Nina explained, was one in which male authority had been paramount, and it was as the patriarch's widow that her mother-in-law indulged in the most intransigent and imperious behaviour. Some time after the old woman's death, Nina and her husband moved to a flat in Venice, but they both found the city too big, unfamiliar and lonely, and they decided to return permanently to Burano, so that she could at last get back to her flowers, her few friends and her favourite nieces and nephews. At the time I knew her, Nina felt the house was too large and empty and she eventually moved to a retirement home in Venice.

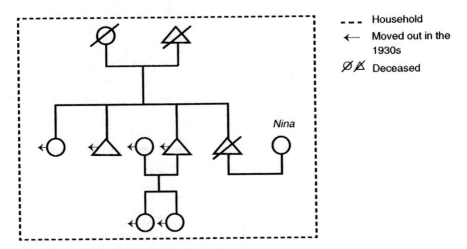

Figure 4.1 Nina's household in the 1920s

While in the past extended family households such as Nina's were not unusual, a mode of residence considered more congenial and desirable was for related nuclear families to live in close proximity. Marta's family was an example of a matrilocal, and mother-centred, grouping, with mother, married son, married daughter and widowed mother's mother living in four very close, although separate, houses. The house of the mother, Maria, at the end of a short terrace, sided on to a canal and its other side was adjacent to the grandmother's. In the next house lived Marta's married brother, while Marta's home was opposite the mother's, across a narrow street.

As in numerous other examples, living arrangements had not been unchanging; Marta's grandmother had lived in that home for about forty years, but before settling there she had moved several times.[2] Both her parents had died in the 1920s, when she was fourteen, and she was left alone in their house with a sister. She already had her *moroso* (fiancé), and as she got pregnant, she joined him in his parents' home. Later she and her husband moved into a home of their own, and they eventually settled in her present house, near her daughter. When Marta married – she too, like her grandmother, was five months' pregnant – her parents gave her a tiny cottage close to their own. After the birth of her second child she moved to a larger house, which she and her husband restored, across the street from her kinswomen.

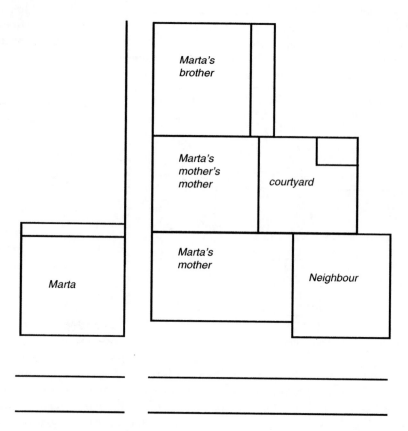

Figure 4.2 Four nuclear families related through women.

According to my informants such groupings were a very common and indeed desirable pattern of residence in the past. In this instance, the three women lived in continuous and intense interaction; they saw each other constantly, and, while they each maintained that they could not possibly share a house and they proudly stated their need and their capacity for independence and autonomy – particularly, as Marta's grandmother said, in the areas of 'kitchen, bathroom and bank account' – at the same time they retrospectively idealised family closeness, affection and solidarity, which they said was being eroded by materialism. Shortly before she died, Marta's grandmother told me that she felt very lonely; she had six living children, but most of them had moved away, they all had their own families. 'They visit me, of course, but that is not the same as if we were *living* together', yet the thought of losing her autonomy and sharing her kitchen was really quite inconceivable (cf. Hirschon 1989: 64–5).

Such deep-seated contradictions and the women's recognition that it was impossible to define an ideal measure of family closeness came fully to the fore when Marta, then about thirty-five, suffered a nervous crisis which she defined as 'mental unease', or 'neurotic illness'. Relations with her family were very difficult, her feelings fluctuating between affection and hostility, and she wondered whether her intense closeness to mother and grandmother might have been one of the causes of her restlessness and misery. In reflecting upon her feelings towards her family members, Marta variously found reasons for her malaise in her relations with her husband, her father and brothers, or else her kinswomen, especially her mother, who she thought may have been the main cause of her troubles.

As for the married brother who lived down the street from her, Marta preferred that relations should not be too close, because he never failed to arouse all her anger and jealousy, sentiments she openly acknowledged and said were due to the fact that both parents consistently favoured their male offspring. Indeed, although by Italian law inheritance should be equally shared, parents often found ways to follow earlier, pre-Napoleonic, tradition and leave larger shares to sons, since it was thought important that real estate should be linked with the family name, which only finds continuity in the male line. As a result, Marta felt that she had been very hard done by.

Her sense of unease was partly due to resentment about her kinswomen's excessive interference, and partly to a fear that her mother may have loved her sons more than she cared for her – her only daughter. Her closeness with her mother and grandmother, mixed with a feeling that they may have been disloyal to her to favour Maria's male offspring, therefore, figured quite prominently as a possible cause of her psychological illness, as it was a constant reminder of her precarious position in the extended family. In an effort to find greater independence, Marta moved to a new house, just a few streets away from her mother's.

Within a few months both Marta's father and grandmother died, and, possibly due to their bereavements, mother and daughter were more than ever together and were frequently seen walking about arm in arm or window-shopping in Venice. Marta told me that, after long reflection, she had understood that her troubles were ultimately due to her gender. Despite her mother's erratic behaviour, even despite her intolerable infatuation with her son (she was 'besotted with him, she thought him a genius, only when he was there did she look secure and relaxed!') Marta had come to the view that her family were not after all worse than others. She and her mother had to share in the common burden of being women. Her troubles were due to the fact of having to live in a world dominated by male pride and archaic traditions – a realisation that ultimately could not fail to reinforce women's mutual affection and solidarity. The problem of whether mothers and daughters should or should not live in great proximity remained unresolved.

Despite Marta's doubts, and her awareness that contemporary tendencies went against such intense family closeness as she had experienced (or, as she said, 'endured') as a young married woman, she still considered that her family's preference for closeness of related nuclei was a fundamentally positive, and indeed a privileged form of residence. Its loss was felt most painfully by her mother, Maria. When I paid her a visit, a short time after the deaths of her mother and husband, I found her sitting alone by her front door in the pale October sun, with an expression of fixed and passive despair. Everything around her, she told me, seemed to disintegrate. As well as mourning her dear ones, she grieved over the break-up of the small row of houses that had long been occupied by her family. Her mother's house, which stood between her own and her son's, had been left jointly to herself and her brothers. As none of them needed it for their own use, it was agreed that her son, who already lived next door, should be given every chance to purchase it. He was asked to name a price, but his offer had been judged too low. When one of the coheirs, Maria's brother, remarked that he would let him keep the house all the same, the young man took that as an insult to his honour and he challenged his uncles to try and obtain a better price elsewhere. Failing to persuade him to accept some compromise, they finally did place the house on the market, and it was rapidly bought by an unrelated family …as Marta and her mother lamented, 'all because of their absurd macho pique and pride!'

Maria then understood that she should have paid her brothers some money without her son's knowledge, but she had not been quick enough to do so, and she could not forgive herself. She could name many good reasons why the house should never have been lost to the family: the first was that it was precisely between her own and her son's home; when her mother lived there, it was just like having one long terrace, but in the future she and her son would have been separated by some complete stranger. What is more, the house contained 'all the memories of a lifetime.', and in it she could still feel her mother's presence. She realised that, despite difficult moments, her own existence, her desire for continuity and her sense of identity were inextricably rooted in the places in which she experienced the greatest intimacy with her relatives, and, above all, her mother. Also, to add to her sorrows, Marta's recent move certainly was a departure from a valued custom. Fortunately, from Maria's point of view, proximity of relations was ensured, given that Marta's cottage had been kept for her seventeen-year-old daughter who had been engaged from the age of fourteen and would soon be married.

Groupings of nuclear families related through men are also present, but by all accounts they were more usual in the past, when fishing was the Buranelli's main occupation and they were particularly widespread among economically strong families of fishermen or fish traders, as we have seen in Nina's example. A preference for new couples to reside virilocally, in or near the household of the husband's father and brothers, forming groupings of nuclear families related through men, was also very pronounced in Venice's agricultural areas of Sant'Erasmo and the Cavallino peninsula,[3] up to the 1950s and 1960s. There too, however, changes in the economy, and, to an even greater extent, an increasing desire for individual affirmation, have led people to establish themselves as independent nuclear families.

Toni and Marisa, a couple I met while on fieldwork, had determined to leave their native Tre Porti and settle in Burano in order to move away from agricultural labour and free themselves from parental authority. For both, life had started in the extended households of their fathers' fathers, while their parents laboured in their respective paternal farmsteads. Marisa assured me that her resolve to have her own separate home and independent economic enterprise, was well-

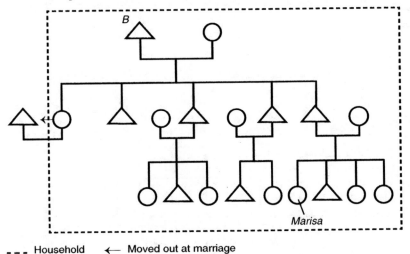

- - - Household   ← Moved out at marriage

Figure 4.3 The household of Marisa's father's father, Busetto, in the 1930s

81

established from her early adolescence. As a child she had greatly enjoyed the bustle and animation of her paternal grandfather's large farmhouse and she had been very sorry to leave. However, on growing up she had soon witnessed a great deal of unhappiness caused by the rivalries and selfishness of household members.

The break-up of her grandfather's household, in the early 1940s, was blamed on the endless fighting among the in-married wives of her father's brothers. Her father, who was the oldest, was the first to leave – a pattern she and her husband decided to avoid altogether, when, on planning their marriage in the late 1950s, they resolved not to join his paternal household, although they were as poor as St. Francis and 'without a share or a trade' (*sensa arte ne parte*).

When Marisa was a child, also her maternal grandparents lived as an extended family with their two sons and two daughters. She was certain that they would have helped if they had been able to, because maternal relations are generally more disposed to be giving and understanding than paternal ones, but in the farming community men had to be hard because it was they who were responsible for all contractual arrangements, cash dealings, sharing of land, and allocation of labour, while women only controlled the production of some vegetables, herbs, flowers and eggs. They were free to sell some of those products for cash, or to give part of them to their daughters and grandchildren.

That household too broke up in the 1930s, soon after the marriages of Marisa's mother's brothers, due to their wives' sexual jealousy. From Marisa's family history, as well as numerous other examples, we see that virilocal extended family residence was widespread in the 1920s and 1930s. Most households, however, tended to split up, although cooperation in farming, especially between fathers

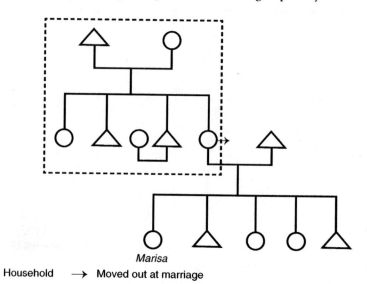

Marisa

- - - Household   → Moved out at marriage

Figure 4.4 The household of Marisa's mother's father in the 1930s

and sons, usually continued, while disagreements were blamed on the women and their litigiousness, especially over children, cash, housework and housespace.

When extended households broke up, young couples sought economic independence as well as a separate residence. A path many followed was to enter some business enterprise and, for inhabitants of agricultural areas, it was a well-established tradition to run a fruit and vegetable shop or a stall at the Rialto market in Venice. Indeed, the vegetable and fruit retail trade is still partly in the hands of a network of interrelated and competitive families from Sant'Erasmo and the Cavallino peninsula. Some of them continue to live on their farms and commute to Rialto every day in the early morning, bringing their produce on large motor boats, while a number of others have moved to Venice.[4]

When, in the 1950s, Marisa and her husband planned their marriage and vowed never to live with either his or her parents, as well as deriving from personal experience, their determination to set up their own household was also consistent with a general tendency for people to move away from agriculture as a full-time occupation, and of a considerable loosening of patriarchal bonds. For Tony and Marisa the price of independence was certainly a high one in terms of hard work and discomfort. A little girl was born, and, after a few years, still dogged by precarious employment and makeshift living arrangements, they decided to open a grocery shop in Burano. Three more children were born, and, despite the fatigue and the constant demands of the business, Marisa said that she enjoyed those years, and even managed to save a small amount of capital.

Rather surprisingly, given their avowed rejection of life in Tre Porti, as soon as they were able to save a small share of their earnings, they started building a small apartment block on land that they jointly owned with their families. When the rented home they occupied in Burano was put up for sale in the 1980s, they still did not feel that they could afford to purchase it, and they decided to move back to Tre Porti where they could at last live on their own property. (Fortunately, they had managed to complete its construction before strict conservation rules brought all building to a halt.) However, as their flat had been let to a cousin, they were forced to evict her, thus creating a chain of housing difficulties, as was not uncommon throughout Italy.

The flat is part of a square and solid, if somewhat anonymous, three-floor building, a type now widespread throughout the Veneto, especially at the outskirts of cities. It is surrounded by a pleasant garden, with tidy flower beds, potted plants, and a painted metal railing and gate. At the time of my fieldwork, the flats were occupied by related families, but, although that may hint at a return to extended family residence, relations were greatly different from those described for the 1930s and 1940s.[5] Each household has its own kitchen, and visiting and interaction are subject to stricter rules and timetables than in the past, given that peoples' claims to privacy are now more articulate and firm. As Marisa lamented, one respect in which tradition is followed is the presence of conflict and competition between siblings.

Although Toni and Marisa's move was precipitated by necessity, it had been long envisaged. However, difficulties soon developed: commuting to work in Burano proved extremely tiring, especially in bad weather. What is more, business was increasingly competitive, because customers had taken to going on shopping expeditions to Mestre and the nearby hinterland, where bulk-buying in the recently built hypermarkets made everything cheaper. Marisa then became absorbed in planning a new business venture: using her proven skills, she would organize the direct sale of organically grown vegetables; this would include all the best local products, much valued by Venetians, and, as well as providing her with a modest income, it would greatly benefit the farmers. With that in mind, she and her husband rapidly transformed and equipped a shed in a nearby field, close to the boat landing. Unfortunately the field was common family property, and, while her older sister and her husband were fairly amenable to this project, her younger siblings felt that their own rights were threatened and they were determined to put every obstacle they could in her way. Her youngest sister was the most determined to stop her. 'Why', Marisa said, 'it is always the young ones, of course, who make trouble! She stayed unmarried to look after the old man, now she thinks everything belongs to her!'

Another difficulty was obtaining a license, but Marisa was confident that that too would have come in time. Meanwhile, a large ground floor room in the house was used as a discotheque for Marisa's teenage daughters and their cousins – something she thought was badly needed, because in the winter Tre Porti had very little to offer in the way of fun and entertainment for the young. At other times, especially before elections, the room was used for political meetings or neighbourhood gatherings (Pink Floyd posters would then be removed very rapidly to be replaced by those picturing Marisa's favourite local politicians). Both Marisa's sons married, and started their domestic life neolocally in their own independent household; one of them moved into a pleasantly renovated cottage in Burano, while the other (whom she pitied, because she thought that, having moved to the vicinity of his wife's family, he would forever be subject to her and her parents' authority) went to live at Ca' Savio, .

Differences in residence patterns between Burano and its nearby agricultural areas are clearly related to their different ecological settings: a preference for virilocal residence, widespread in the Cavallino peninsula until the 1950s, was clearly due to the fact that for farmers, who usually rented their land from absent landowners, intensive cultivation and cooperation between agnates were of the utmost importance for economic survival.

By contrast, in Burano, with its densely built urban structure, the ideal form of residence is, and always has been, neolocal, but having relatives as neighbours is considered extremely desirable, and the strongest preference is usually for closeness of mothers and daughters. Most young married women will avoid living with either set of parents, but, should circumstance force them to do so, it is sharing with a husband's mother that is truly regarded as an almost certain source of trouble, since antagonism between mothers- and their daughters-in law is generally strong, and is emphasised in much folk literature, sayings and proverbs[6] (although some husbands' mothers do live in harmony with their sons' wives, particularly when the latter have a job).

The matrifocal residence pattern of Marta's family, which I have described above, follows a long-standing tradition. As informants explained, it certainly made good sense, especially in the past when men were often away fishing, that kinswomen should want to live close together in order to share events such as childbirth or illness, to cope with loneliness, and to bear their anxiety over their husbands' safety at sea. However, matrifocal residence is by no means the rule; as we have seen, groupings of nuclear households linked through men are, or used to be, considered desirable. For example, in the past among economically strong families of fishermen, boat builders and traders residential choices were dictated by the economics and organisation of work. In particular, fishing was traditionally based on the cooperation of brothers under the leadership of their father or another senior kinsman, because it was a firm belief that to achieve full efficiency and safety at sea crew membership should be based on common blood (cf. Vianello 1998).

In sum, while preference for matrifocal residence is possibly the more widespread, its converse is also present, both at an ideological and a practical level. The examples I have introduced, in addition to wider observation and data, mainly lead me to comment on the great versatility and resilience of household arrangements, which clearly recall the broad original meaning of *familia*, and the fact that availability of resources is a very strong factor in determining choices.

Kinship terminology

Burano's kinship terms are, on the whole, very similar to Venetian ones, as differences mainly concern accent, pronunciation and the extent to which somewhat old-fashioned terms, such as *mare, pare, amia* and *nessa* are used, especially to designate, rather than address, relations. For example, young people will use *mama, papà* when speaking to their parents, but they often use *mi mare, mi pare* when they speak about them. We thus have *mama, papà, zia, zio,* which are the same in present-day Venetian and Italian, in place of *mare, pare, amia* and *barba,* but the latter are used habitually by older people.

Kinship is bilateral, but subtle nuances in the use of kinship terms show that, traditionally, agnatic emphasis was rather pronounced and that, as for Venice in general, patrilineal kinship was legally more binding than matrilateral relations, especially from the point of view of inheritance and patrimonial succession. Indeed, sibling ties are given different weight according to whether they are derived from father, mother or both.

Biases or inclinations I observed during fieldwork are also detailed in old dialect dictionaries and juridical manuals (Boerio 1850; Ferro 1780). There are,

Table 4.1 Kinship terms

Venice and Burano	Italian equivalents
M: *mare, mama*	*madre, mamma*
F: *pare, papà*	*padre, papà*
S: *fio*	*figlio*
D: *fia*	*figlia*
MM, FM: *nona, ava (less frequent)*	*nonna*
MF, FF: *nono, avo*	*nonno*
MB, FB: *barba, zio*	*zio*
MZ,FZ: *amia, zia*	*zia*
Z: *sorèa*	*sorella*
B: *fradèo*	*fratello*
FBD, FZD, MBD, MZD: *zermana*	*germana*
FBS, FZS, MBS, MZS: *zerman*	*germano*
ZS, BS, SS and DS: *nevodo*	*nipote*
ZD, BD, SD, DD: *nessa*	*nipote*

Abbreviations are: M for mother, P for father, S for son, D for daughter, Z for sister B for brother; these are combined where necessary, for example, MM designates mother's mother; FM, father's mother; FZD, father's sister's daughhter; MBD, mother's brother's daughter and so forth.

for example, different degrees of brotherhood, in diminishing order. The terms for 'brother/sister', *fradèo, sorèa,* mean 'born of the same parents', full brother/sister, and are glossed in Italian as *fratello/sorella carnale,* or *germano(a)*. *Fradèo/sorèa bon(a)* (good), or *de sangue* (by blood), also defined as 'legitimate', designates a brother/sister born of the same father and different mother. *Fradèlo de mare,* or *fradelastro,* where *fradèlo* has been suffixed with the pejorative ending *astro,* like *sorelastra,* half-sister, designates the relationship of siblings who have the same mother but different fathers. Uterine siblingship, however, as the dictionaries concede, could be legitimate if the siblings were born within successive legitimate marriages. The third degree of siblingship, described by the phrase *fradèo/sorèa bastardo/a,* and glossed in Italian as 'fratello o sorella naturale', designates the relationship of siblings born of different illegal unions. The weakest sibling relation is that derived from the fact of two people having had the same wet nurse. The terms are *fradèo/sorèa de late,* rather inadequately rendered in English as 'foster brother/sister' (Devoto 1962).

In present-day usage, the term *zerman* can be rather confusing: it is often used for 'cousin' in general, but its original meaning is that of father's son, especially in the phrase *fratello germano,* that is, either full brother or patrilateral half-brother, or, as in *cugino germano,* parallel patrilateral cousin, that is, father's brother's son.

Table 4.2 Affinal terms

Buranello	Italian
HM, WM: *madòna*	*suocera*
HF, WF: *missier*	*suocero*
SW: *niòra*	*nuora*
DH: *zènero*	*genero*
ZH, HB: *cugná*	*cognato*
WZ, BW: *cugnada,*	*cognata*

Abbreviations are: HM, WM, husband's/wife's mother; HF, WF, husband's/wife's father; SW, son's wife; DH, daughter's husband; ZH, HB, sister's husband/husband's brother; WZ, BW, wife's sister/brother's wife.

In ordinary conversation, as distinct from legal usage, *zerman/a* is often preceded by the second person possessive pronoun (yours) and it thus subtly indicates that the speaker dissociates from the relations of the person he is addressing. For example, marriage partners will occasionally refer to one another's kin as *'to mare'*, *'pare'*, *'fradèo'*, *'sorèa'*, *your* mother, father, brother, sister, to distance themselves and withhold the respect and affection those terms would normally imply; it is, in other words, a slight undermining of the affinal link with an interlocutor's blood relations and a way of reaffirming the boundaries between families related through marriage, sometimes momentarily, but at other times quite permanently through repetition.

Such emphasis on different strengths and degrees of sibling relations clearly draws our attention to underlying notions of the physical basis of kinship and, above all, the significance of blood and of ideas about conception.

The terms which differ most from Italian, and, in particular, from southern Italian dialects, are those which designate a spouse's parents. Indeed, the terms traditionally used in central and southern Italy, which are now part of the national language, are *suocero*, *suocera*, from the Latin *socer*, or *socrus*, derived from an Indo-European root which, in origin, indicated exclusively the husband's parents, and the fact of a person belonging to a given family group (hence also *soror*, *sodalis* and *socius*).[7]

Madòna, Italian *madonna*, which derives from the Latin *Domina Mea*, is equivalent to 'my lady'. In the southern and central Italian vernaculars, however, *madonna* was never used to designate or address a spouse's mother as it is in various northern Italian dialects. In the Middle Ages, *madonna* was used mainly as a title of respect for women of high condition, or, in the poetry of the *Dolce Stil Nuovo*, for the loved woman. Its use is very common in early Italian texts, and is in fact earlier than its devotional use as a term for 'Mother of Christ'. Religious overtones, which are, in any case, later than the simple hierarchical use of the word, are far removed, even quite absent from the speakers' consciousness. Indeed, very few of my informants actually pointed out connections between *madona* as mother-in-law and as the Holy Virgin, except in a spirit of humour and parody, as, for example, when they quoted the saying *la Madòna sta ben 'n`tei quadri* , 'the Madonna is best in pictures' (like the Virgin Mary), where the pun is clearly based on the superimposition of the word's two different meanings. On the other hand, the phrase *dona e madona,* meaning absolute mistress – an expression often used in early notarial documents and contracts, and which has survived in everyday language – derives solely from *madona* as 'mistress', that is, 'my lady' in its hierarchical sense.

While *madòna* is most frequently used as a term of reference, some people prefer to address their wives' or husbands' mother by its modern near-equivalent, *signora*. That usage too, however, is perceived as one that signals embarrassment and avoidance. It may be considered appropriate to a transitional stage in the early days of marriage, as some spouse's parents – and especially mothers – hope eventually to be addressed as *mamma* and *papà*, while young men and women find some difficulty in doing so. Much embarrassment is saved in Burano's egalitarian atmosphere, given that most people start from infancy to address adults by their first name, while, at the same time, respect is shown by the use of the third person singular pronoun, *ea, eo* (Italian *lei*) and not the second person *tu*, for a husband's or wife's parents.

The male equivalent of *madòna, missièr* (in Italian, *messere,* or, for prelates, *monsignore*) is similarly a title that grants superiority, and which in Venice was sometimes given to important persons like the Doge, the Procurators of St. Mark's, or a ship's first captain, who would be described or addressed as *Missièr Grando*. It was sometimes also given to saints, and in particular to St Mark, but clearly it never acquired religious associations comparable to its female equivalent *madona*. In the kinship sphere it was also used in the phrase *missièr pare* – a form of address that went out of fashion in the late eighteenth and early nineteenth century, although *missièr/missie,* by itself is still commonly used in Burano for father-in-law.

Fiol/fia, nono/nona, barba and, more rarely, *amia* are also used in a classificatory way either to designate or address, respectively, younger or older persons in a familiar and sometimes affectionate way. *Fia* and *fio* are applied to apprentices, errand girls and boys, pupils or anyone junior in age and skill. *Ava/avo* are sometimes used ironically to underline the great age of the person concerned by the archaic quality of the word itself. *Barba* and *amia* also connote an element of respect, which is not entirely absent from *nono/nona*, but while the latter connote compassion and sympathy as well as deference for old age, *barba* and *amia* convey a feeling of admiration and recognition of the person's adult status and superior competence. In the past they were commonly used to address instructors and teachers, but, as well as respect, *amia* and *barba* also connote trust and affection. St. Antony of Padua, as Marta explained to me, is a perfect representation of avuncular tenderness. With lilies in the background, and the baby Jesus in his arms, he seems to symbolise the male equivalent of motherly love, and indeed, when she prays in church she often looks up to his image as much as to that of Mary.

Except for the instances I have discussed, we may observe that Italian, Venetian and Buranello usage largely overlap. It is, however, Italian that shows the greater readiness to adopt new, and often foreign, terms of address. For example, the English 'mummy' signals an attempt to change and adopt new and more 'modern' attitudes in the mother/child relation, but such mannerisms, now sometimes accepted in the cities, would be regarded as utterly ridiculous affectations in a small community such as Burano.

Ritual Kinship

The general terms for wedding and baptismal sponsors are *compare* and *comare*. However, a distinction is made between a wedding and a confirmation or baptismal sponsor. The first, who assists the bridegroom and bride during the wedding, is described as *compàre de l'aneo*, godfather 'of the ring', the second as *compàre de San Zuane*, or godfather of St. John. *Sàntolo/sàntola*, then designate a confirmation sponsor, as well as baptismal godmother and godfather, while the sponsored child is addressed or described as *fiozo/fioza*. *Compare* is used also for comrade, age-mate or friend, while the feminine *comàre*, as well as close friend, also means 'old wife', midwife and gossip.

The phrase *fia/fio d'anima* indicates informal adoption of an orphaned or very poor child into a family or into the household of a single person. Thus, it describes any caring friendship towards a younger person. The tie is mainly conceived as a charitable one, but one notoriously open to exploitation. At present its mention is often ironic, as the kind of patronage implied is clearly out of favour.

Both marriage sponsorship and godparenthood, however, are important relationships in Burano (R. Parkin 1997: 124–5). Contrary to eighteenth- and possibly nineteenth-century custom, few people today seek to form ties of ritual kinship with socially superior outsiders. Wedding sponsors are often a couple's closest and most congenial friends or relations, while at present godfathers are often chosen among relatives – a practice supported by the Church, and which reflects the need not to prejudice potential alliances, because, while god-parenthood, *comparaggio*, may be superimposed on the kinship relations, the reverse is against canon law, and would thus lead to the further narrowing of an already limited and complex marriage market.

Choice of a senior family member is a sign of trust and respect, and both marriage and godparenthood, which are valued features of Burano's social life, generally involve people in rather expensive, but highly enjoyable, exchanges of gifts. Being asked to act as *sàntolo* or *compàre* is then regarded as an honourable and welcome opportunity for a person to show his/her social worth, although it is sometimes looked upon as a burdensome and demanding financial commitment.

The Physical Bases of Kinship

Notions of gender, socialising practices and jural rules are naturally linked with ideas about the physical bases of kinship: flesh, blood, semen, milk and sometimes tears, are therefore powerful and all-pervasive natural symbols (Douglas, M. 1970). For Buranelli, as for Italians in general, blood is the main ingredient and binding substance of kinship, it is the basis for true unity and solidarity. It symbolises the soul, or at any rate a spiritual substance that transcends the individual and keeps a descent line together. However, blood is also the seat of the passions; a fright can turn blood to water or ice, while anger will cause it to heat and rise to the head, therefore it is somewhat feared, especially in men. By an interesting convergence of folk beliefs with Aristotelian traditions,[8] men's blood is generally held to be warmer than women's, indeed, heat and blood are two important elements in the symbolism of manliness. For example, a show of temper, which is evidence of 'hot blood', although it may be feared, is not disapproved of in men, since it inspires respect.

While both male and female substance are thought to contribute to the formation and identity of persons, male and female blood are thought to have different weights in determining the strength and quality of kinship, and in particular of its jural rules. Although women are said to age and lose their fertility earlier than men, it is nonetheless thought advisable for men not to be profligate in their sexual activity: as the saying goes, 'after 50, keep your sap for the tree' (*dopo i cinquanta, tien el sugo per la pianta*). Indeed, the male emission of sperm is considered equivalent to the loss of forty times as much blood – and that is, according to my informants, the cause of postcoital exhaustion and melancholy. It follows logically that male-derived kinship should be held the stronger and the more legally binding. In that respect, folk notions resemble those expounded in learned treatises, in which sperm is described as the refined or 'distilled' product of men's blood, or, as Vico writes, the 'flower of blood' (1968 [1744]: 263), while breast milk is its more diluted derivative, often described as 'coagulated blood', and, therefore, it gives rise to the weakest of all sibling relations, that of 'milk brotherhood'. However, just because of its total absence of jural force, milk brotherhood, the link between persons who were fed by the same wet nurse, has been much romanticised and is often used as a metaphor for 'brotherhood' of all men, outside the sphere of legal and patrimonial links.

Sexual activity, conception, gestation and childbirth are symbolically linked through metaphor and analogy with the lagoon environment. The woman/moon association is particularly strong, and women frequently illustrate descriptions of their bodily states with tidal images and with notions of flux, circulation, flooding and so forth, which are clearly derived from observation of the rhythms, or feared excesses, of the lagoon. For example, some of my informants maintained that removal of the ovaries would cause a woman headaches, because, putting an unnatural end to menstruation, it would 'block everything up' and hinder the free circulation and renewal of blood. Like the moon and the weather, women are said to be changeable and subject to sudden turns and moods, 'woman and moon today clear, tomorrow dark', '*dona e luna ancúo serena e doman bruna*'. Moreover, since human reproductive rhythms are analogically related to the phases of the moon, the tides and the movements of fish, conception, like successful fishing, is most likely to take place when the moon is new, full, or waning, for at those times, tidal variation is strong, inflowing waters (*de vegnua* or *in colma)* are abundant, and the fish 'mount' from the sea into the lagoon.

An image frequently used to describe the way a new person is formed, and one that recalls beliefs that led Carlo Ginzburg's Anabaptist miller to fall foul of the Inquisition, is that of the mixing and leavening of bread (1976; Ott 1981: 191–197). For example, Piero's accounts of occasions that, he believed, had led to the birth of his numerous offspring, usually ended with 'and so, that is how we kneaded the dough'. As we have seen, the foetus is thought to be formed from both maternal and paternal blood, which get mixed during the first three months of gestation. In the following three months the bones develop, and the flesh is formed during the last three months, which are often described as the 'feeding phase'.

Gender differences begin to make themselves felt well before birth, as they give rise to differences in both the foetus's and the mother's state of health, especially in the final months of pregnancy: a female foetus may cause weakness and lassitude, as well as leg pains, because of a tendency to put all its weight on one side of the mother's body. By contrast, male foetuses, although they do not move and kick in the womb as much as females do, are thought to be well formed at an earlier stage than are females, and after the first three months they communicate to the mother a sense of energy and well-being that enables her to sustain high levels of activity and enjoy all sorts of heavy labour. If she suffers from nose-bleeds, or if, in an advanced state of pregnancy, she starts white-washing her house, it is almost certain that she will bear a boy. What is more, if at the time of conception, or during pregnancy, she was affected by some inflammation, that would certainly disappear with the delivery.

Socialisation, Gender and Change

Emphasis on gender begins very early in life: when a baby is born a ribbon of the appropriate colour, pink for girls and blue for boys, is hung at the door of the parents' home, and its sex is immediately signaled by the colour of clothes, while relatives and friends will pass complimentary remarks on the fact that a child's appearance clearly shows all the marks of her/his sex at a very early stage.

Respect and obedience to parents, great caution in relations with persons outside the family and emphasis on honour and shame (which will be the topic of my next chapter) are often present in Buranelli's discourses and in socialising practices. However, in the 1980s parental authority was by no means as rigid and forcefully enjoined as it was in the 1950s and 1960s. Indeed, it was noticeable that people who were themselves the parents of teenage or young adult children often expressed a sense of revulsion against the narrow strictness, and sometimes the violence, that characterised their upbringing.

Memories of deprived and unhappy childhoods and descriptions of punitive and tyrannical fathers were not unusual, but, comparing the narratives of people who had grown up in agricultural areas with those of numerous Buranelli, it appears that farmers were traditionally the most demanding, strict and sometimes cruel, of fathers. Buranelli's explanation, no doubt tinged with slight antagonism, was that peasants in general had a stronger tendency to subject their offspring to toil in order to accumulate and to save any money surplus, with which they were eventually able to buy some of the lands on which they previously worked as tenants. They said that fishermen were in any case at the mercy of providence and fortune, and less given to 'capitalising'.

Burano's senior men were on the whole better disposed towards cooperation with the young than towards the relentless exercise of authority. At present, the number of professional fishermen has gone down considerably and fishing is often exercised as a sport or a part-time occupation, but when it was the sole form of employment, obedience was a practical necessity due to the nature of the enterprise, rather than to any tyrannical propensities on the part of fathers and senior men. Instructions had to be obeyed efficiently and quickly, and orders were delivered in ways an outsider might have thought rather brusque and imperious, but it was in everyone's interest that they should be carried out without questioning, because speed was essential for successful fishing, and in situations of sudden bad weather delay might have put the men's lives at risk.

Although some elderly fishermen remembered that they were brought up to regard their father with a respect that was almost reverential, and although they were well aware of the hardships of a fisherman's life, they recalled that, as children, they were very eager to be taken on fishing expeditions and enter their trade. At the time of my fieldwork, a teacher complained about absenteeism from school and about the fact that explanations such as 'going out fishing with father' or 'helping prepare the boat', 'unloading', and so forth, were thought to be legitimate justifications for failure to attend lessons, or reason enough to leave school in the middle of the morning. Teaching their sons the family trade from a very early age, therefore, generally led to tenderness and attachment, rather than antagonism.

Exceptions are not unknown. When, on a cool October evening, a man ordered his son to deck out their boat, the latter refused point blank, saying that he simply did not like to go fishing at dusk (*de ponente*). It was one of those clear evenings when, after days of sirocco, the water is wavy and the fish come near the

surface, but the young man had already planned to join his friends on an outing.
So, when his father repeated his order to change back into his working clothes,
the young man walked out and, in a furious temper, he set the boat on fire. He
then went parading round the High Street and Piazza.[9] Like other acts of filial
rebellion and instances of father/son hostility, the episode was highly dramatised
by the two men involved and followed by their friends and neighbours with
genuine interest. The main damage, as people commented, would probably have
been suffered by their insurance company, but curiosity was mostly focused on
the two men's relationship, and people wondered how long it would take for them
to be reconciled, take up work together, and start fighting again. Indeed hostility
between men for whom cooperation is an economic necessity is always regarded
as a ruinous eccentricity to be avoided at any cost.

Generational differences, however, emerged from some of the older men's
criticisms of the young and their repeated complaint there was 'no respect any
more', often followed by descriptions of the hardships they had suffered in their
childhood and youth, before Italy's economic miracle had reached the Venetian
islands. The main change for Burano's fishermen had been the introduction of
engines, which relieved them of exhausting rowing and sailing. No less important
was the introduction of nylon nets, far less time-consuming and costly than
cotton nets, which required constant cleaning and repair. An increased demand
had greatly raised the prices of fish, while new methods, such as the use of
dynamite and of turbulent power-operated rakes could produce large yields very
rapidly. Such methods were eventually forbidden, because they damaged both the
lagoon's soil and the fish stock, but they had enriched a few families.

Older people were the first to acknowledge that they had greatly benefited
from their improved economy and better living conditions, but many of them
lamented a loss of traditional values, which they saw as an essential part of their
culture and a prerequisite of their trade. As one man observed,

> Now they are all millionaires, they are always nice and clean, and all they do the
> minute they come home from work is go bragging and boasting in the Piazza. But they
> are not half as clever as we were, because it is their instruments that are doing the
> thinking. We used only our heads!

For men who had grown up before, during or soon after the Second World
War, learning had usually started between the ages of eight and ten. This was not
due only to their great poverty, but also to a firm conviction that to become truly
competent and develop a fully adult masculine identity, boys have to start
learning their trade very early – a view fishermen shared with Venetian craftsmen,
since, as they say, 'learning is a life-long process' and 'experience deserves respect'.

At the beginning of a boy's apprenticeship, whether he was working alongside
his father, grandfather and uncles or under another master, his first tasks were to
carry the senior man's pipe and coat, keep the boat dry by sweeping up any water
that might have collected in it with a small wooden pan (or *sèssola*) and shine its

metals with oil and sand. If he was taken on a long trip, he would be in charge of washing up, airing the men's clothes, and keeping an eye on the nets when the men were having a rest. His first reward was usually a good meal and a few coins, but his pay would gradually increase as he acquired more skills. By the time he was fifteen or sixteen he might begin to earn a fairer share, but he would not be considered to be fully adult until he returned from his military service at the age of twenty-one or twenty-two, and only then would he begin to take part in fishing as a full partner.

By the time a boy began to earn his wages and started taking girls out and drinking in pubs, he might resent any purely social obligations towards his mother and sisters, and avoid going on too many family outings. When, on Sunday visits to grandparents or relatives, sons were prevailed upon to join the family, it was usually the women who presided, served food and generally managed the situation and defined its character. Conflict would then be controlled and expressed mainly through teasing and joking, while the presence of members of the extended family made it easier both to air and to contain some resentment that may have been seething within the nuclear family.

When men went to one of the piazza's cafés after work, they generally stopped to chat with members of their peer group. In public places – usually thought of as 'male' – men who may have been the most tender of fathers and playful of grandparents in the home setting would sometimes find it a trial to have to adopt a social manner towards their own sons. For example, if a son turned up where his father was drinking with friends, both might experience a slight embarrassment or sense of incongruity. When such situations occurred, however, awkwardness would be overcome with the father's show of surprise and pride at the chance of offering his son hospitality in the form of a glass of wine, which the young man would quickly drink up, before leaving the premises.

During their long apprenticeship, boys were encouraged to sharpen their spirit of observation and 'to steal with their eyes' any new method or technical innovation (cf. Herzfeld 1997). When, in the old days, lagoon fishing was conducted without even a compass, they had to learn some astronomical notions, and relate the phases of the moon to the rhythms of tides and the behaviour of fish. They had to become familiar with the winds and predict weather changes. Above all, to orient themselves in poor conditions of visibility, for example, when the fogs descended suddenly and blotted out all familiar landmarks, they needed to resort to their own mental map of the lagoon, and it greatly helped to know how long it takes to cover certain distances, taking account of the force of winds and the direction of currents.

Another essential quality, especially for younger crew-members was a capacity for secrecy. If they were lucky enough to find some rich fishing ground, Buranelli would never speak of it to anyone other than family or partners. On the contrary, they might give wrong directions and misleading information. Narratives about prodigious catches in some obscure location were usually told by old men reminiscing, long after they bore any practical relevance. As Vianello points out in her study of Pellestrina (1998: 56–7), Burano's southern counterpart, a fisherman's contest with his prey is not so much one of strength as of cunning, *furberia*. In describing his pursuit of a large school of mullets a man spoke of their amazing deceitfulness; 'they know that we are after them, they *understand*...they can be very tricky, but we have to win in the end!' An example is that of the fishing of *gò*, perhaps the humblest of preys, but one greatly valued, especially at times of scarcity at the end of winter. To describe someone as a fisherman of *gò* was really to class him as the most humble of men – a fisherman without a proper boat, and one whose activity was somewhat more like collecting than hunting. Small, rather ugly, of a colour as brown as that of the mud in which it hides and almost impossible to enjoy because of its many soft bones, the *go* is generally viewed with condescension. Nevertheless, the ability to catch *go* is often mentioned as a typical example of Buranelli's subtle skills and ingenuity.

Unlike other fish, which migrate to the sea in the cold winter months, *go* dwell all year round in the warm muds and shallows of the lagoon. When its reproductive season approaches, in early March, the males begin to prepare their dens, which are like minute narrow tunnels, where the females will lay their eggs. As the saying goes, 'at the time of San Joseph (19 March), when the ink fish enters the lagoon, the *go* prepares the nuptial bedroom to welcome its female' (*De sant'Isepo la sepa monta e el go fa la conca*). As the dens are progressively widened, the eggs, which resemble tiny clusters of grapes, are left hanging from their roofs, while the females of the species wander off, leaving the males to guard them, and it is at this time that the chase intensifies.

As Piero recalled, for him *go* fishing had been 'lesson one', a rather lengthy and unpleasant procedure that mainly consisted in tying one or more traps to a thin pole fixed to the soil, with their openings against the current. Sometimes his traps would fill within half an hour, but other times not one fish would enter them. Rather than return home empty-handed, he would get off the boat and try to catch a few *go* in shallow waters, searching for them under the mud with his naked arm, a practice the mere memory of which still caused him to shiver with cold.[10] In addition to the 'trap' and 'naked-arm' methods described by Piero, *go* could be caught with a fish-hook and cane, or hunted with the use of a tiny harpoon. All such practices, except for cane fishing, which Buranelli told me is the least efficient, were sometimes discouraged, because it was thought that they might put the *go*'s reproduction at risk.[11]

Against those who feared its extinction the fishermen argued that they knew very well how to feel at the touch if the fish was a pregnant female, and they would then set it free, in order not to deplete the lagoon of its resources. Such skill, foresight and sensitivity, they maintained, was completely lost to the young, who lived for the moment and had little concern for the future, 'they destroy the soil of the lagoon by dragging their heavy nets for miles, while their heavy power-operated rakes for gathering clams are killing all its vegetation. Their use in the lagoon was forbidden, and substituted with vibrating rakes, but it is not at all sure that they cause less damage.'

The senior men's complaint that there was 'no respect' any more did not refer only to a lack of respect for age and experience, but also to an absence of adequate understanding and care for the environment. In their view, modernity had brought about the loss of certain qualities, which, although specifically developed in the context of their fishing culture, are conceived of as bearing strong moral values that transcend a simple and crude pragmatism. Their loss, it was implied, would be a great loss to the community, because they were the very basis of its ethos and identity. Courage, discretion, patience, a willingness to learn and to make sacrifices for the sake of their trade, a capacity to think logically and act decisively and quickly were all summed up in the Buranelli's notion of *intelligence* – a quality they associated mainly with masculinity.

It was partly the result of a long fishing tradition (although at present occupations are quite diverse) that gender roles were strongly differentiated and fixed at an early age, and were in many ways different from those of agricultural areas. For example, in Sant'Erasmo women have traditionally been very active in the retail vegetable trade, they are expert rowers, and they are among the most frequent winners in women's regattas. By contrast, in Burano till the 1980s anything to do with boats, their construction and upkeep, as well as their use in fishing and transport, was the exclusive concern of men, and women did not do any rowing, neither did they participate in the fish trade nor even the cleaning and preparation of fish for marketing.

Father/daughter relationships were radically different from those of fathers and sons, but what was sometimes construed as the preference of a father for his boys may have been just the outcome of the division of labour, through which fathers shared a greater range of interests, and inevitably spent more time with sons than they did with daughters, rather than to a real difference in love and affection. By the age of about ten or eleven, girls seemed to have wholly entered feminine roles and to have developed behaviour traits that closely imitated those of their mothers and older sisters. Domestic tasks, 'helping', are learnt very early and are usually acknowledged and rewarded. As one of Burano's school teachers said to me rather unkindly, 'they always want to help, they are like little maid servants'. As the local school only goes up to age fourteen, if a girl's parents decided that she should continue her education, she would have had to travel to Venice every day, at considerable expense and with some discomfort. She would have to rise early, and return home well after one o'clock in the afternoon, missing the midday meal, which is in general a valued moment in Italian family life. To make things worse, Burano's children sometimes found Venetian schools quite forbidding; they were teased about their accent, and, in some instances, about their poor performance, which was mostly due to the inadequate teaching in their local secondary school.

As a result, at the time of my fieldwork, few girls had completed their upper school education, although a small number had done so, and two young women were away at university. Some girls had started to attend upper secondary school courses in Venice, but after one or two years they had tired of commuting, and they decided to employ their time in ways they considered more profitable, that is, going to work, so as to save up for their future, and, in some instances, to help their families. School discipline was, in any case, resented, and many of the subjects were considered extremely boring and irrelevant to the practicalities of the adult working life that they were impatient to enter.

An important aspect of girls' socialisation, and one which usually began in earliest infancy, concerned the fulfillment of their gender roles and the development of domestic and affective skills. By the time a girl was three or four, parents as well as relatives and friends would start making remarks about her femininity. These were usually made in affectionate admiration; they were mainly acknowledgments of the girl's attractiveness, sometimes meant to bring about the thing they stated, but they were not entirely without a hint of ambiguity, for they might have implied that the child was innately sensual, and that her instinctive seductiveness would help her to prosper in a competitive world. However, such compliments were an indirect acknowledgment that it is a sinful world in which such female wiles are desirable. Indeed, the problem of infantile sexuality certainly remained unresolved in Burano's native psychology, and one was sometimes led to wonder whether the white angel-like overalls of Italian nursery school children, and the frequent references of nuns to their innocence, may not have been due to a will to negate and repress something that the secular population are dubious and puzzled about, and which they sometimes gloss with some confusing oxymoron, such as 'innocent malice'. Girls, it was often said, are naturally coquettish, they invite playful teasing and admiration, but they certainly must be watched. It is, thus, a rather dubious compliment to say to a parent, albeit in banter, 'you are going to have to watch her!' (*questa ti darà del filo da torcere*, 'she will give you plenty of yarn to twist', in other words, trouble).

As girls grew up, fathers' gradually became more distant and reserved. By the age of about twelve or fourteen, girls had learnt to take orders from their father, since respect and obedience were more or less taken for granted, but the main burden of upbringing was usually left to mothers. Given that people are strongly aware of their complex and multiple kinship relations, a fact that narrows the number of permitted alliances, early marital choices were approved, and it was not at all unusual for girls to get engaged while they were still at secondary school. After an initial period of reserve, fathers usually gave their consent, as it then seemed that the natural sensuality and cunning (*furberia*) shown by the girl in early infancy had worked to a good purpose, and she was expected to settle down in marriage – never to be a threat to their peace of mind and respectability – except for the period of engagement, when the relationship might break down.

Marital commitments usually came to be by the initiative of the couple concerned, and the extent to which families acknowledged and celebrated them was a matter of choice, and was also dictated by the extent to which the match was approved. The usual expression in dialect to describe a steady relationship or assiduous courtship is *aver el moroso/a.* Courtship sometimes began when a young man started walking a girl home from school or work, and gradually took charge of her in a variety of ways, especially making it clear to other potential suitors that she was not available. However, to have a *moroso/a,* which literally means 'amorous' (while 'lover', *amante,* would have had the wrong connotation, given that *morosi* are not supposed to have sexual intercourse, although they may actually do so) may mean different things, and may describe different degrees of intimacy, from an incipient friendship and a passing infatuation to a permanent commitment.

Until the early 1960s, it was considered necessary for a young couple who wished to get to know one another well without incurring censure from their families and neighbours to make the nature of their relationship known, and, although that is no longer obligatory, it is still fairly usual for girls in Burano to give their betrothal some recognition, with a small celebration and with an exchange of gifts. Therefore, in the 1980s, some families marked their children's impending or potential engagement by having a meal together, although their relations were often characterised by frequent disclaimers and by the fear of breakdown. ('If they are roses, they will flower'). Those who were more attached to the Church would attend Sunday Mass together; they would then exchange a present – usually a plain ring (*fedina*), but more recently a stone ring for the girl and some personal object for the young man. In the afternoon they might go on an excursion to Jesolo or Tre Porti with their parents, and sometimes grandparents. They might begin to go about by themselves or with other young couples when the relationship was well established.

While an early marital choice was usually approved, and was a pattern most girls envisaged as normal, it was certainly not a path that every girl was able to follow, or wished to conform to. As women got older, the range of possible behaviours became quite diverse, given that they would not necessarily conduct their lives solely within the narrow confines of family, village and Lace School. While early marriages were certainly approved from a sentimental point of view, as well as from a purely demographic one, they were sometimes frowned upon from an economic viewpoint, because they left little chance for the partners to save enough money to buy a home and properly set it up with furniture and linen. A fear that children might arrive very rapidly, thus curtailing a period in which young couples should have prepared for a well-organised entry into adult married life, was often used as an argument to persuade them to wait and not rush into marriage.

Since the 1960s and 1970s it has been quite usual for women to go to work, but for the majority of schoolgirls, marrying, having children and making a good home were the dominant ambitions. Only two out of about forty girls I knew wanted to be doctors. One of them was, in fact, planning to go to medical school, while for the second, who was still at secondary school, medicine was tied up with marital as well as religious fantasies; she told me she would marry a famous surgeon, work with him at an African mission and have four children whom she would leave in Burano in the care of her parents. Fairly popular choices were nursing or accountancy, and professional aspirations were usually traditional ones, for example, becoming hairdressers, shop assistants or shopkeepers, factory workers or clerks, while numerous girls, as well as married women, worked in Murano's glass factories, checking, cleaning and packing the glass. As women worked and became financially independent, parental authority weakened and emphasis on honour and respectability sometimes became no more than a source of tension and squabbling, which may have acted as a brake, but did not substantially govern a woman's behaviour.

Relations between brothers and sisters were usually close during childhood, but often became distant after marriage. Although brothers sometimes had to bear some financial or moral responsibility, especially for unmarried sisters, relationships between siblings of different sex were sometimes characterised by antagonism and particularly in matters of inheritance they were often marked by hostility, so that sisters might end up being regarded as falling, rather awkwardly, between the categories of kin and affine.

Mothers and sons were often very close, and their interactions cooperative and affectionate. Mothers advised their sons, and took their sides in conflicts with fathers. For example, it was usually mothers who supported and encouraged sons who chose a long education. Their relationship thus might become increasingly close over the years – sometimes a mother's *revanche* against the husband's male world, if she had found marriage isolating and difficult. Not all mothers, however, were ready to go through long years of austerity to support their sons' ambitions. Several men complained to me that their mothers had been too strict, especially in demanding that they should go to work at a very young age, and some adult men were still bitter about having been forced to start their apprenticeships at Murano's glass factories at twelve or thirteen. 'You only get your mother's respect when you bring home the pay packet.' Indeed, financial security and ease are guaranteed when at least two members of a family bring home their wages.

Although most mothers were deeply involved in their offspring's marital choices, they did not usually attempt to control their sons' sexuality. On the contrary, as they explained to me, they would be more successful in delaying their sons getting married by ignoring, or even mildly encouraging, a certain amount of sexual freedom in their twenties and thirties. For example, one woman assured me that she would have been very happy to see her youngest son married with a nice respectable girl. For the time being, however, she was happier to have him at home.

> He can have a wife for a month or two when he goes on holiday with his friends to Latin America. He did that last year, then he sent this woman a large Christmas present; I must admit I was rather jealous, but he said: 'remember what it was like when *we* were hungry?' What can I say? He is my youngest – not a sullen and hard type like his father at all. We get on very well.

Since the young man was the last of her offspring who still lived at home, it was arranged that she would leave him her house, and he was thinking that they should go to a notary together, so she could write a regular will, or his sisters and brothers would all come to claim their share, and one sixth of her cottage would really be no good for anyone.

While many people claimed that a double standard is natural, as they hold it for certain that male sexual drives are far more compelling than female ones, it would not be correct to think that men were always encouraged to develop, or show, aggressively male sexual traits and behaviour; indeed, chastity before marriage was valued in men too, both by mothers and by wives, not only because it was reassuring, but also because it was thought a virtue, associated with honesty and goodness.

In conversations about family sentiments and the exercise of authority, Buranelli often affirmed that also dead parents and grandparents are very much present in peoples' minds. Regular visits to the cemetery in Mazzorbo were, for some people, part of their daily routine. A number of elderly widows used to visit the graves of their husbands every day, and they regarded tidying up and arranging some flowers as part of their household duties which had been such a dominant discipline throughout their lives. Some people have the distinction of a family grave, obviously more expensive than an ordinary grave in the earth, while in general there is a strong preference for burial in the walls of the cemetery which have been raised to considerable height to cope with a lack of space.

The dead, especially those who are in Purgatory, sometimes appear in peoples' dreams; at moments of great danger at sea, they are invoked by sailors who feel particularly close to them, and they sometimes come to remind their relatives to pray and celebrate masses for them, so they may help them overcome their difficult period of transition and penance. Sudden and sometimes frightening apparitions of dead relatives, however, may also occur quite unexpectedly, when people are awake during the daytime. This is particularly the case with persons who died suddenly, and whose souls are still full of resentment. A number of episodes I was told occurred soon after the First World War.

On All Souls day, the graveyard, as the women say, looks like a festive drawing room. Mass is celebrated there while most families cluster round the graves of their relatives. Women divide their attention between the graves of their husbands' and those of their parental families. The evening before, however, most households, and especially those of people who have recently been bereaved, are in some suspense. Some people claim that they can hear noises and loud banging as if someone were moving furniture about, and even voices, especially in the attics. It used to be the tradition to leave a place laid at the table for those who had recently departed. Over a large white napkin there would be left a jug of red wine with a glass and some bread beside it; sometimes a plate of broad-bean soup or other food would be left covered by a napkin, as was usually done when a member of the family was expected to return late from work. In some instances, walnuts, which are harvested in late October and early November, would also be left to be enjoyed with the wine. For most people the scene I have described is a childhood memory, but, according to my informants, some families still do this, or, if they feel self-conscious about showing that they are superstitious, they just 'forget' a flask of wine and a glass on the table.[12]

In the old days, fishermen would be very reluctant to go out fishing on All Souls, for, as one of them related, 'our old people were so nervous, that if they just saw a tiny sea bass jump into the air, they thought it might be the ghost of one of the departed'. They would keep a lamp alight on the prow of their boats, and they particularly feared the bites of poisonous fish, which were thought to be the souls of the damned. Such vague fears were also widespread in agricultural areas, one inhabitant of Tre Porti told me that, going back home one dark evening around the beginning of November, he suddenly felt that the air had gone very cold, he heard a heart-rending cry, then a huge black dog appeared near a ditch between the dirt road and the lagoon, soon to disappear again, while its howl followed him all the way home.

Reflecting upon such beliefs, people often attributed them to 'the past', or 'old folk'. Or else, at the same time as they related instances of supernatural events as their own personal experiences, they often stated 'I don't actually believe that the dead can come back...but...' Such open contradictions, however, may themselves be a way in which they express a suspension of their disbelief, and they underline the fact that they are entering ground uncharted by either the Church or by some other reassuring authority. Reports of such events may, despite their disclaimers, be a way of conceding that 'who knows, there may be something in it after all'. What is interesting and puzzling here is the transition from naive credence to a sort of disbelief, which still allows, or in fact leads to, a desire simply to tell the story. But, whether or not people believe that such stories contain any literal truth, what is mainly emphasised is the fact that family sentiments, and even power relations, last well beyond death and the grave.

A Concern with Endogamy

Most Buranelli described their attachment to the family and their keen awareness of kinship networks and links as sources of solidarity. However, living amongst people they could almost universally regard as relations also gave rise to some apprehension about the potentially negative outcome of long-term inbreeding – a fear strongly fueled by outside medical personnel, teachers and priests. For numerous informants, the exercise of recording the names and kinship degrees of their antecedents was not at all new, for, due to their long-standing care to avoid very close unions, they had usually done this with the village priest when they were planning their marriage. An added difficulty was the discrepancy between Church and state reckoning: Italian civil laws only prohibit marriages between direct ascendants and descendants by blood, filiation or adoption, aunt and nephew, or uncle and niece, while Canon Law also prohibits marriage between first cousins, who, by Catholic reckoning, are related in the second degree. Until 1917 prohibitions extended to third cousin range. As appears from patterns of marriage in Venice's islands, marriages between cousins were common in the 1920s and 1930s, but, as they were discouraged both by the Church and by medical opinion, they were often surrounded by an aura of anxiety.

A good illustration is that of a woman, Francesca, whose surname, Bianchi, is a very common one (see section on nicknames). As she explained, while trying to reconstruct her genealogy for me, this surname occurs in her family with brain-tangling frequency, both among her consanguines and her affinal relations. 'They are all Bianchi.' That is a great inconvenience. 'When you start courting somebody who might be too close and therefore forbidden by the Church, you go to the priest and he makes an inquiry called *proava*, which covers seven generations.' Both her parents had the surname Bianchi. Her father's parents were also both Bianchi. Francesca knew that her mother's father was not a Bianchi (in the diagram, y) but she did not know either his surname (that is, her mother's maiden name) or the surname of her mother's mother, because she only remembered them by their nicknames – for all she knew, her mother's mother too might have been a Bianchi.

Francesca's husband's paternal grandparents were also both called Bianchi, while his mother's father bore a different surname, and the surname of his mother's mother was unknown to her. Leaving aside the possible confusion due to the frequent occurrence of the same family names, what the woman knew, and what was in fact most relevant to her degree of relatedness with her husband, was that her own and her husband's fathers were the children of brothers. She and her husband were second cousins – a degree of relation for which dispensation was easily granted.

Given their low population figures, inhabitants of the neighbouring areas of Sant'Erasmo and Tre Porti were equally concerned with fears of inbreeding. Thus, for example, Marisa, who at the time of my fieldwork was the mother of two adult and two teenage children, still felt somewhat uneasy about her degree of kinship with her husband. Her misgivings over endogamy begin to be substantiated when we look at her and Toni's genealogies over time. There we see that marriages between members of their respective families have taken place in subsequent generations. Thus, if we consider both sets of grandparents, we see that by Italian common thinking Marisa and Toni are second cousins, although, as the woman recalled, the village priest 'who had worked it all out' had 'wrongly' reckoned that they were third cousins. 'That wouldn't even have been so bad', Marisa said gloomily, 'if there hadn't been another person…a certain Romana'. For, as she explained, although there is no kin relation between her husband's parents, the fact that, in Marisa's own words, both were her father's cousins made her and Toni 'not only second cousins, but second cousins twice over'.

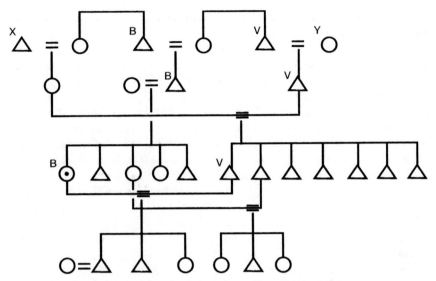

Figure 4.5 Marisa's kindred. B = Busetto; V = Viani; X and Y = other

As we see from the diagram, Marisa's paternal grandfather and Toni's maternal grandmother were brother and sister, while her paternal grandmother was the sister of Toni's paternal grandfather. Although Marisa showed strong animosity against her husband's mother, whom she referred to as 'that woman' or 'the *redoubtable* Romana', it was a long time before she openly acknowledged that the woman was both her mother-in-law and her father's cousin.[13] As well as a long-standing clash of characters and interests, Marisa's angry references to Romana implied a wish to suppress the fact that the woman was, like herself, the daughter of a Busetto married to a Viani – a wish that might have been more satisfactorily assuaged had her culture provided a ritual outlet, such as, for instance, the Nuer breaking of kinship, rather than the priest's acquiescence (Evans-Pritchard 1951: 31–6). In some alpine villages, when cousins got engaged, they would 'burn away the kinship' by holding a lighted candle under the palms of their hands, but I found no memory of comparable rituals in the Venice area.

In reality, Marisa was well aware of the bilateral nature of heredity, but she found computations very cumbersome, so that she never reached any precise numerical picture of the 'biological relationship coefficient' in her marriage, or any realistic idea of the genetic risks involved (if any); she just obscurely knew that the overlap between her blood and Toni's was a little too great for comfort, and she still resented the fact that, when she was young, she must have been given either too little, or contradictory, information by her parents, her priest and her doctor.[14] However, neither her concern nor her parents' disapproval of the marriage had been strong enough to prevent another Viani-Busetto alliance, that between her sister and her husband's younger brother. But when somebody suggested that her sister's daughter would make a perfect match for her son, Marisa just averted her face in anger.

Two more points should be made to end this section: the first concerns the ramifications of kinship through different areas of the northern lagoon and the second concerns relations between different households. As we have seen in my chapter on history, Buranelli tend to emphasise their difference from people living in Venice or in other islands. Looking at some of the kinship diagrams that I have collected, however, we see that families, in fact, extend and ramify throughout the northern lagoon and sometimes into Venice and its hinterland. For example, to follow a family with which we are already familiar, Marisa's kindred spreads out through Tre Porti and Burano, with one or two affinal links in Torcello, and several (not included in the diagram) in the Venetian hinterland and the historical centre.

Thus, despite Buranelli's often stated preference for marriage within the island community, which may explain a tendency towards early romantic commitments, spouses are often found outside Burano and kinship networks are quite extended, so that one would be hard put to find an entirely local kindred or genealogy of more than two or three generations.

Looking at relations between members of the different households, both within and outside their kindred, I found that, while some of the persons I knew were bound to one another by a variety of common interests, exchanges and so forth, others barely knew one another by sight. Elements of choice and personal interest were important factors in determining a person's social circle and the range of most frequent and active interaction, as I hope to show in my section on friendship and social life.

Marisa knew all the women who bought food at her shop, addressed them all by their first names, and was in almost daily contact with a few old people she sometimes gave food to, or helped in a variety of ways. She knew my landlady, whom she sometimes came to visit while I was her guest, but, as she told me on a later visit, their familiarity lapsed after I left Burano. She knew only a small number of people in Terranova or Giudecca; for example, she did not know Marta at all, and was only vaguely acquainted with her family, although they were in fact connected through a distant affinal link.[15] While she was strongly identified both by herself and by others as a native of Tre Porti, Marisa was actually related to a large and prosperous Buranello family, as a sister of her husband's father had married and come to live in Burano in the early 1930s, while two more aunts must have come to live in that island as well, either through marriage or for some other reason, and are now buried in the graveyard in Mazzorbo.

Tracing all kinship links between different households would generate a most complicated entanglement and criss-crossing of lines, but what I have written so far may be sufficient evidence to show that the islanders' statement that they are 'all one family' is not merely a metaphorical one and certainly is not without some objective basis.

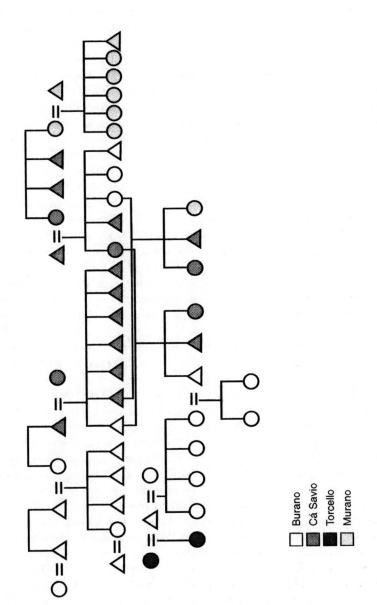

Figure 4.6 Members of a family spread through different islands

Burano
Cá Savio
Torcello
Murano

Names, Surnames and Nicknames

Surnames, in Italian *cognomi*, are inherited from the father. Traditionally, women take the family names of their husbands at marriage, but a change in the law in 1976 now permits them to retain their father's surname alone, if they so choose. First names, usually given when a child is baptised, or when a birth is officially registered, are very often those of a grandparent, either in the maternal or paternal line, while, as an alternative, or sometimes in addition, a child may also bear the name of her/his baptismal sponsor. Having a child named after one is considered a sign of devotion and respect, because it implicitly shows its parents' desire that the child may develop and perpetuate some of the qualities of the person thus honoured. As David Sutton observed during his fieldwork on Kalymnos (Greece), the giving of names may raise serious conflict between families, for naming is thought to establish continuity through the generations, but an important feature of the naming system is also its connection with the transmission of property, because 'the child who takes the name of a grandparent (or other relative) will inherit the property of that grandparent (1997: 423; cf. C. Stewart 1991: 58).[16]

In order to satisfy conflicting expectations, some children may bear several names. In some instances, and partly to avoid repetition and to break with tradition, a child may be given an additional 'new' name, that is, one not derived from either grandparents or *compàri*. In that case the traditional name or names are written in her documents, but the child is generally addressed and known by the new name.

Given that Buranelli traditionally favour the names of the island's patron saints, the pool of available first names is rather limited. The reason, according to a linguistic study of medieval naming (Folena 1971: 445–84) is that in Venetia both names and surnames were fixed in written records considerably earlier than in other Italian regions, between the eleventh and the twelfth centuries. As a result, very few names of German origin or of vulgar and chivalric derivation were added to an early repertoire of the Greek and Latin names of a narrow circle of local saints' of great religious prestige and diffusion, especially at the popular level. The Venetians' attachment to their cults, as well as the strong cohesion of kinship groups and the singular weight of tradition, therefore, favoured a concentration first of personal and then of family names.[17]

With altered attitudes to family solidarity, and under new political conditions, there developed new categories; numerous surnames derived from trades and professions, like Scarpa, Ballarin, Busetto, Rizzo, Bullo, Calzavara, Tagliapietra, Favaretto, Ferro and so forth, while ethnic and geographical ones, like Trevisan, D'Este, Pavan, Boscolo, Padoan, Veronese, which possibly reflect long forgotten displacements and migrations, became widespread from the fourteenth and fifteenth centuries.

Also many nicknames, once they were written in documents, and thereby made permanent, were eventually treated as surnames. Nicknaming is thus not entirely different from the giving of surnames, except in so far as its practice preceded the spread of literacy. The interesting question, then, is why such an archaic form of naming should have continued in Burano (or indeed other areas) alongside the more conventional manner of naming favoured, or even imposed, by the Church and the bureaucracy. In the first place, as my informants maintained, nicknames fulfilled a very useful function in separating branches of an original family.[18] Viewed in the light of relations between centre and periphery (be it Burano in respect to Venice, or, as in this case, Venice in respect to Italy), the explanation may also lie in the islanders' resistance to an order imposed from the outside, and an affirmation of their separateness and of their preference for an 'in-language' which they need not share with others.

The islanders' account of the way in which nicknames came to be designated as *detti* (singular, *detto*) may go some way towards supporting that view, at the same time showing their versatility in absorbing standard Italian words into their current vocabulary and adapting them in their own idiosyncratic way. In this case *detto,* the past participle of the verb *dire,* to say, which in its ordinary usage simply means 'said', has become a substantive, and is used as a synonym of 'nickname'. As nicknaming stubbornly continued long after the introduction of records, and given the confusingly high occurrence of the same names and surnames, the phrase *detto x* is often included in early Venetian documents, like, for example, *Decime,* Inquisitions trials and notarial records. It was also increasingly entered in government documents and reports after the unification of Italy (1866), when police, *carabinieri,* schoolteachers and bureaucrats, who were mostly recruited from other Italian regions, found it very difficult to identify people by name and surname.

Confronted with the population's obstinacy in refusing to respond to various government summonses, requests for tax payments or orders to report for army service, in exasperation officials always added 'said' followed by nicknames in personal documents and in the city's registers. Buranelli themselves then adopted the Italian *detto* for nickname in place of their earlier *soranome.* Such continued use of nicknames is an aspect of the liveliness of the dialect in the entire Venetian region, especially in the remoter areas of the countryside, and in the city's popular quarters, Cannaregio, Castello and Giudecca.

Surnames too, as has emerged from a comparative analysis of Italy's regional telephone directories, have retained the highest incidence of dialect forms. As we have seen, they were codified in the local vernacular between the eleventh and twelfth centuries, when Italian was relatively undeveloped. In more recent times, after Venice became part of the Italian state in 1866, Venetians made little effort to italianise their names, given that their dialect enjoyed the prestige of a language: it was traditionally spoken and written by the ruling class as well as the people, and it therefore naturally reflected the city's complex social structure by its different registers, accents and styles that clearly connote distinctions of education and social class (Lepschy et al 1996: 70).

107

Despite the Romantics' association of antiquity and closeness to Latin with nobility, and despite observations on the unique musical appeal of the Buranelli's long-lingering vowel sounds, their dialect, like their naming customs, is now commonly regarded as uneducated and rude. All the same, many islanders show remarkable inventiveness both in their speech and in their introduction of new nicknames. Their explanation for the continued use of nicknames is that surnames are few and that, as families grow larger, distinctions have to be made between their different branches. Moreover, even families not related through kinship sometimes bear the same surnames, especially those derived from place names or trades. Nicknames are then more specific than surnames and they are often used to keep distinct people who might have a common ancestor, but whose kinship is, in fact, rather distant.

Most nicknames last about three or four generations, while family names could, at least in principle, last in perpetuity and include wider and wider circles of cousins. Consequently, in Burano, where people are ever aware of the potential ill-effects of inbreeding, nicknames serve the useful purpose of symbolically 'splitting kinship', in order to remove any potential embarrassment when someone marries a person who bears the same surname. Use of nicknames to designate different branches of one original family thus fulfills a useful cognitive and classifying function, and, although kinship reckoning for the purpose of marriage is usually based on surnames and Church records,[19] some people maintain that nicknames have a positive psychological effect, and they are often invoked to disclaim kinship and to make useful distinctions between collateral descent lines.

An example that I have already discussed above is that of the surname Bianchi. Not unlike its English equivalents, White, Whites or Whitehead, this is fairly widespread throughout the country. Indeed, not all the families who bear those names are related, and some instances of homonymy might be fortuitous, or might be due to alliances between unrelated Bianchi. At the same time, however, also those who are joined through some blood link, as well as by surname, are subdivided into smaller groups through the use of nicknames.

As I mentioned in my discussion of marriage, in describing her kindred to me, one of my informants recalled the nicknames of each one of its members, as well as the way in which they were transmitted through three generations; she bears the nickname of her father, *Sepa*, which he inherited from his mother (not his father). Her two sons bear her own nickname. Her husband's nickname, *Folpo*, was passed down in the male line from his father's father, but her husband's five siblings took over the nickname of their mother, who was nicknamed after her father, *Baicolo*.

Here then we see that transmission of nicknames is randomised in contrast to that of surnames, and gives rise to new groupings. While surnames are passed down in the male line, nicknames can be inherited either from the mother or from the father. This, according to some informants, is because they are based on

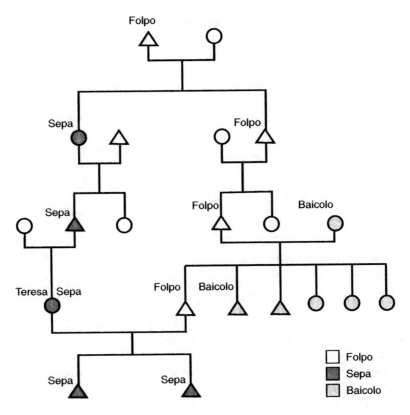

Figure 4.7 The transmission of nicknames.

resemblance so that a child will usually bear the nickname of the parent whose features or character she or he most closely recalls. Beyond physical resemblance, it is also a matter of social incorporation, as when the mother's family of origin is larger and better known, or better off than the father's, and the mother's father is particularly inclined to absorb one or all of his daughter's offspring into his sphere of activity and protection: the child is then usually referred to by the nickname of his mother, which may be one passed down from her father or mother. As a result, not only patrilateral cousins, but sometimes even brothers, can have different nicknames.

In some instances, a woman's *personal* nickname is transmitted to her children and grandchildren, giving rise to a matrilineal dynasty. What is more, while women often refer to children or young adults by the nicknames of the mothers, men are more likely to identify them through the father – a fact probably due to the separation of genders, through which the social world is ordered in notionally distinct male and female spheres. Indeed, flexibility and freedom of choice in the use of nicknames bears interesting relevance to Buranelli's actual experience of kinship: like their residential patterns, it shows greater openness than the patrilateral ideology and jural tradition of the dominant Venetian, and Italian, culture.

While the transmission of surnames is inevitably formal and rule-bound, nicknaming is innovative and subject to improvisation. Nicknames usually given first to individuals, may be passed down for two, three or four generations – a number more or less equivalent to the islanders' genealogical memory. Some nicknames, however, go back a long way. For example, in account books and wage registers dating from 1872, now in the archives of the Lace School, workers are designated both by surname and nickname. Some nicknames are the same as those in use today, while others, mainly personal rather than family ones, are now lost to memory. At the time of my fieldwork nicknames that had acquired particular resonance were those of families or individuals with special claims to fame, such as those of rowing champions, who are noted and admired for their prominence as representatives of the island in competitive boat races against other islands or Venetian neighbourhoods.

Anthropologists have studied nicknaming customs from different viewpoints. Cohen, Mewett and Strathern see them mainly as expressions of 'belonging'. Mewett emphasises the way in which they are used by insiders, so that they symbolically mark the boundaries of a community. As he writes in his article on a Scottish crofting village (Mewett 1982: 238), 'through nicknames, people can express a collective view of the social identity and character of each individual in the community'. Anthony Cohen makes a related point when he suggests that each individual may be likened to a piece of a jigsaw puzzle, in that 'each piece stands in a given and determinate relation [to others] such that, when one is missing, or incorrectly placed, the character of the whole is affected' (Cohen 1978: 449. Quoted in Mewett 1982: 238). Here, then, nicknaming is viewed as essentially positive since it reinforces feelings of community and group identity.

By contrast, Gilmore describes nicknaming in a Spanish village as a cruel and distasteful expression of backwardness and as a manifestation of extremely negative and destructive impulses. As he writes, 'Hostile nicknaming, and an extreme touchiness about them, seem to flourish in particular types of Mediterranean communities', and especially those 'characterised by structural familism and psychological atomism' in which, 'individual autonomy, honour and male assertiveness have been institutionalised and internalised as primary values'. Gilmore's hypothesis is that nicknames are 'a form of verbal aggression, a displacement of competitive envy and sexual, economic and status competition, especially among men' (1982: 687–700).

Quoting Freud, he asserts that, like satirical jokes, comedy or wit, 'they give intense pleasure to those who use them', and this stems from the successful release of repressed feelings or wishes in disguised form; in this case, it is safe to say hostility (695). 'Nicknames then represent a kind of sadistic comic relief, a displacement for repressed antagonism, for animus...' They are, 'powerfully felt threats to the very integrity of the person'. They can be, 'an attack not only on self-esteem, identity and family honour, but also on manhood itself: a kind of symbolic castration'. They 'demean', 'depersonalise' and symbolise antagonism against families as well as individuals (697–98).

Also the nicknames of competitive Sarakatsan shepherds described by Campbell are often derogatory, but they can be laudatory and underline a person's strength and good qualities (1964: 300–315). In his study of a village in the Spanish Sierra (1971 [1954]: 160–169), Pitt-Rivers points out their function as a form of social control. More recently Brandes (1975: 141) explores their classificatory use, he too, however, describes Spanish villagers as feeling 'an intense abhorrence of nicknames, as well as the belief that they are backward and degrading', so that it is avoidance, rather than use, of nickname address that can solidify social ties with friends and relatives. 'After all,' as his informants observe, 'you never hear of a doctor or a lawyer or a big city person having such a name'. He too describes nicknames as mainly concerning men. Both Gilmore and Brandes, after initiating their discussion mainly with reference to their fieldwork areas, end by extending their statements to 'Mediterranean rural life' and 'Mediterranean societies'.

Much evidence from Burano, however, shows that attitudes to nicknaming are far more nuanced and varied than those of the Spanish villagers studied by Gilmore and Brandes. In the first place, it concerns women as well as men; moreover, and especially in the past, nicknames were not only used by Venice's uneducated and by the poor, but were also applied to members of all social classes; indeed, particularly vivid and scurrilous ones were in use among the city's aristocrats.[20] In so far as dialect speaking in general is denigrated, also nicknaming, which is typically a vernacular customs, is sometimes looked down on, particularly by outsiders, whose attitudes are at the same time suffered, resented and occasionally echoed by the islanders.[21] At the time of my fieldwork, opinion was divided; as we saw above, accounts of nicknaming were offered to me spontaneously (in itself a sign of detachment) and, while some young people regarded the custom with a degree of condescension or self-consciousness, as an expression of the more archaic and rude aspects of their culture – one that might have brought derision to individual persons, or given rise to the contempt and disapproval of their community – most people were quite indifferent to it and took it entirely for granted. Others simply acknowledged that nicknames are useful in respect to kinship and social relations, as well as (and not least) thanks to their potential amusement value, and they spoke of nicknaming with a degree of affection, particularly when they associated the nicknames with some specific occasion on which they were acquired, whether in their own or their parents' or grandparents' time.

As they explained to me, nicknames are usually applied in an informal and haphazard way, and they are often an expression of contingencies, such as a person's mood, a flash of wit, or a mocking intention. Sometimes they are unwittingly self-inflicted, as when a tired or frightened man exclaimed, 'I am dead!' ('*Ah mi so morto!*'), and 'dead man' (*Morto*) became his permanent nickname, or when another complained in self-pity, 'they have crucified me', ('*I me ga meso in croxe!*') and he and his family were thereafter known as 'cross' (*croxe*, a common metaphor for 'a pain'). Other nicknames reflect a special aptitude, or refer to a trade, the name of a fish, a tool, or any other creature or object, or they may record some memorably comical incident.[22]

Some Buranelli, anticipating that the custom may eventually be abandoned and that some of the nicknames may be altogether lost to memory, had made a record of all those they knew. A written list had been compiled to serve a purely practical purpose, that is, the delivery of gas containers. Until 1984, most family cookers had been operated from such containers distributed from a shop in the main street whose owner was the sole agent responsible for their provision and prompt delivery. To keep track of customers' accounts and addresses (and avoid the potential confusion due to the small number of family names, and the fact that women would often require new gas supplies in a great hurry, because they only noticed that it was running out in the middle of cooking) each customer was registered with her/his first name, nickname and address. When gas was piped in, and the register of nicknames was no longer needed, it was suggested that it should be kept in perpetuity, since it would eventually become a true 'monument' to Burano's folk culture – a view that illustrates the islanders' feelings that change was impending. At about the same time, a comical reading of all nicknames was recorded, and was loudly broadcast for the villagers' amusement on a Carnival evening. The performance was a great success, and, as was reported to me, people were highly amused, some of them doubled over with laughter, and certainly no offense was taken.

It might be argued that Buranelli's attempts to exaggerate and ridicule aspects of their own culture are certainly a way of 'removing' them, relegating them to the past, while at the same time taking care to commit them to memory. It is as if the islanders, only too well aware of having traditionally been the butt of others' ridicule and caricature, had decided to be the authors of their self-parody. With greater experience of national and urban life than most of them could master in the past, they thus see some of their customs in a highly comical key, and by celebrating nicknaming at Carnival, a time of symbolic reversal, it was as if today's modern and enriched Buranelli had entered a confrontation with their own culture, which they considered archaic and, in some ways, rude and naïve, and which they themselves were beginning to classify as 'folksy'.

Whether nicknames will continue to be used in the way that they have been until recently is quite unpredictable. Some people thought that they may not, but schoolchildren did very often attach appellations, both to their friends and to those they disliked, to a much greater extent than in other areas. This was sometimes done in fun and was readily reciprocated, but sometimes (and often in vain) it was bitterly resented and opposed. At the time of my fieldwork a nine year old girl, Chiara, had started calling one of her school mates *gatto*, 'cat'. This had provoked such anger that the victim's mother had gone to confront the offending girl's mother, as well as the schoolteacher, to ask them to stop Chiara calling her daughter a cat. When they tried to reassure the victim's mother, saying that it was only a childish prank that would have no consequence, she replied, showing genuine concern, 'Oh, no; it will stick. I know it is going to stick'.

'Anyway, what is wrong with "cat"?' asked the mother of the offending girl. To which the first woman replied firmly that it was not a matter of right or wrong, 'We are called *Canarin* (Canary-bird) and my daughter does not want to be called "cat". If your girl does it once more, I promise you I shan't come to protest on my own. I warned you'. Questioned about the reason for calling the other 'cat', the girl explained, 'She scratches. She has scratched me more than once'. There was a kind of inevitability about naming her schoolmate 'cat', and, while to my knowledge the punitive visit threatened by the girl's mother did not take place, the two families continued to be resentful and remonstrations were made every time the two mothers met by chance in the street.

Thus, despite a feeling that the nicknaming custom may soon come to an end, schoolchildren still attach nicknames to their mates – or their little enemies – far more frequently than in Venice's schools, and the matter may lead to quarrels between adult members of the families involved, as there is always a fear that such unwanted appellations may become permanent. In general, however, nicknames that acquire a strong degree of permanence are mainly those generated in the adult peer group. Whether they are laudatory, neutral, or scurrilous and offensive, the best attitude for the recipient is one of indifference and humour. As for the proud Cretan shepherds described by Herzfeld (1985: 234–237), complaints would be childish and a capacity to rise above even the most obscene of nicknames is a test of real character. In so far as the battle involved is one between an individual and society, it is one that individuals generally lose, but in that case they must make a show of moral superiority and nonchalance.

Notes

1. At present the Church actively supports parental authority.
2. 'Forty years', 'fifty years', 'a lifetime' are not always to be taken literally.
3. The litoral of Cavallino, about 10km from Punta Sabbioni to the old Piave, is almost entirely taken over by camping sites and hotels. Beyond the bridge over the Piave Vecchia the peninsula is connected by road to Punta Sabbioni and Jesolo. Its name, *Cavallino*, 'small horse', may be derived from the Latin *Equilio*, since the area was originally rich with forests and horses.
4. A position at the market is considered very desirable. Licenses are not easy to obtain, and, in some instances, they have been kept in the same family for generations (*de riedo en riedo*, from succession to succession). Some men, at the time of my fieldwork, worked as clerks or handymen in Venice's offices, but, although such positions offered a regular income and social security, pay was very low, so there was a need to supplement their earnings by cultivating some land for cash, or by additional, and sometimes unofficial, employment.

5. Due to patterns of ownership, tenancy laws and difficulties in finding domestic help, clustering of nuclear families is now widespread also among Venice's bourgeoisie and among the owners of large palaces that can be divided into independent units.

6. In a poetic dialogue, or *contrasto*, between two women in Buranello dialect, Ma' Passerina complains to the mother of a girl who has rejected her son's marriage proposal, and enumerates all his virtues. The girl's mother answers, 'it is not *he* she is worried about, it is you, crabby old woman; she does not want to become a drudge in your home!' (Ateneo Veneto 1998: 39).

7. Compared with *madona* and *missier*, *suocera* and *suocero* may imply a fuller incorporation of the in-marrying wife in the husband's family. *Madona* and *missier* connote notions of hierarchically structured, but essentially *social* relations.

8. 'The female sex is dominated by the moon, which always 'looks to the sun and follows it' (Aristotle, *De Generatione Animalium*).

9. An easily irate and uncontrolled person is called a *cazza-fogo*. Anger gets a man heated, or literally causes him to catch fire. In exasperation, he may set fire to objects, like the home, or, as in this case, the boat that was the locus of his father's authority.

10. *Go* often figure in conversation, as well as in Burano's cuisine; their habits are well known and are often referred to in relation to the seasons.

11. In the eighteenth century it was forbidden to capture *go* between the first day of Lent, or about mid-March, and the end of June.

12. A custom still widespread in parts of rural Veneto is to offer and eat small, multicoloured macaroons, called *fave* or *fave dei morti*, where *dei morti*, 'of the dead', refers to the season when they are consumed. Sweet *fave*, described in old cookery books as 'hard and shapeless biscuits' and also called 'dead man's bones', *ossi da morto*, are a sweet version of fava beans, which are widely associated with death and used in divination

13. That link was not evident because Romana's connection was through her mother, and she therefore bore a different surname. A folk belief that heredity is transmitted more powerfully in the male than the female line, can generate confusion.

14. Patterns of kinship and marriage strategies in peasant societies are systematicaly analyzed in Jacquard and Segalen 1973 and Segalen 1986. Concern with inbreeding, a major theme in Italian nineteenth-century novels (Sciama 1977) was sometimes present in 1980s Burano. As Segalen points out, however, 'Even human groups conscious of a cultural originality they wish to preserve by means of marriage between blood relatives need only a few partners from outside the group to renew their genetic patrimony completely (1986: 122).

15. Marisa has no kinship connection with Antonia or with Anna, but she and Anna have a common affinal link.

16. As Sutton writes, exploring Kalymnian naming strategies helped him to understand their sense of 'continuity, both spiritual and material' (1997: 421). The link between names and property explains Greek debates over the name Macedonia for the former Yugoslav republic. By contributing to our understanding of nationalism from the bottom up, his study of naming shows the relevance of kinship analyses in Western nation states (see also Herzfeld 1992: 68).

17. At present, saints' names, like Bartolomeo, Basilio, Emiliano, Gregorio, Oriano, Quirino, Valerio, Vitaliano, have a minor place, especially in comparison with other Italian regions (de Felice, E. 1986: 14). However, Albino remains widespread in Burano.

18. Like Latin *cognomina*, nicknames are usually added to first names (*prae-nomina*) and surnames (*nomina*) to introduce a distinction between families descended form the same *gens*.

19. According to a sceptic, prohibitions were invented so that parsons might be able to charge the poor for granting dispensations (cf. Goody 1983: 134–46).

20. Whether they were first applied to them by their equals, or their servants and dependents, the nicknames of aristocratic families eventually entered common usage because they served a useful classificatory function. Some derived from architectural features of their houses, for example Barbarigo *dalla Terrazza*, Corner *dalla Ca' Grande*, Dona' *dalle Torreselle;* others from

some marked characteristics, either physical or psychological, such as Corner *Pampalugo* (stupid, gawky) or Contarini *Fisolo* (bird of the marshes), while coarse ones, such as Gritty *Mona* (female genitals, fool), Pisani *Merda* (excrement) and Dolfin *Culatta* (rump), are certainly not lacking. In these instances too, over time semantic content becomes secondary, while the nicknames remain as useful labelling and classificatory devices.

21. The dialect, which has preserved customary forms of naming and address, shows remarkable continuity and is often preferred by persons who are perfectly capable of speaking Italian. The practical advantages of speaking *the language* are evident to all, but a deep attachment to vernacular culture remains – as is most effectively shown by some of Italy's leading poets, among them Pasolini and Zanzotto. Many Buranelli resort to a kind of bilingualism, perhaps better described as diglossia (A. Lepsky, G. Lepsky & Voghera 1996).

22. Other examples are: *Grazia* (in the sense of plenty), *Pancin, Ciapate, Bello, Caenazzo* (lock), *Strigheta* (little witch), *Volega* (a type of fishing net), *Galiotto, Carestia* (penury), *Spuin, Cuccagna, Marinaro, Batteletto, Fra Lusso, Padrin, Buriello, Rama.*

5
STRATIFICATION

Equality needs no reasons, only inequality does so.

(Isaiah Berlin, *Conceps and Categories*)

As Piero once observed, 'what a pity we have to die, now we all have enough to eat!' Like many others who suffered privation in their early years, he viewed all Buranelli as having had the same, or a very similar, start in life – usually a very precarious one. His view showed an interesting coincidence with those of outsiders who described Burano as an island of paupers, with no differentiation, at the margins of the city.

By contrast, people who had been more fortunate would usually talk of poverty as the condition of others. For example, the women whose housing vicissitudes I reported below, each in their own way expressed a strong awareness of social differences and statuses. Nina drew a sense of security from her husband's prosperous position. She designated social categories as *ceti*, a word which, with its old-fashioned implication of 'orders', does not carry the same connotations as 'classes', which, in the current political climate, she regarded as more confrontational.[1] Marisa worked very hard to ensure that her sons should obtain white-collar jobs, while Marta, conscious of some inherited superior standing, regarded any dependence on the social services as demeaning. They too spoke of the past as a time when poverty had leveled out hierarchies and had brought about privations also for the better-off families. Such memories lent validity to their conceptualisation of their community as one imbued with egalitarian values, which, however, were contradicted by human tendencies to emulate and compete with others. In their view, new possibilities for gain and social change after the Second World War had greatly eroded an earlier spirit of equality. At the same time they had allowed for many of the islanders' bourgeois vocations to fully develop and create new differences, especially in terms of consumption and of investment in education for the generation then growing up.

When probed further, Piero too would admit that, even in the past, at times of greatest poverty, differences always existed. There have always been the haves and the have-nots, *siori e puareti,* those who never stopped eating and those who starved.[2] The fact that some families are, in reality, richer than others was due, Piero thought, to a combination of skill and luck, since economic choices often have unpredictable outcomes and no amount of hard work and intelligence could lead anywhere without good fortune. For example, no one could have foreseen that a small number of local families would have been quite as successful as they have been in taking over the lace trade on an international scale. 'When the boys [at the time of my fieldwork middle-aged men] were little, they would count themselves very lucky if their father came home with a pound of *go* for their supper.'

While Piero's emphasis on equality was not representative of the views of all Buranelli, it showed the coexistence of different and contradictory visions; one based on an egalitarian spirit, through which differences were considered inessential in fundamental notions of personhood, and the other derived from an acknowledgment of inequalities and a spirit of emulation.[3] Such apparently inconsistent descriptions of social structure were inextricably tied up with peoples' experience and understanding of the past: as for other aspects of Buranelli's identity, the past was a touchstone and a basis for enduring ideologies, but it inevitably implied a dividing line between 'then' and 'now'. By all accounts, the most significant improvements in Burano's economy have taken place since the Second World War, while its impoverishment was most severe from the latter part of the eigteenth century, and, as we shall see in my chapter on lacemaking, it became very grave when, in 1872, weather conditions made fishing impossible.

A rough comparison between census data on Burano's population and trades at the turn of the eighteenth century with those of 1981 may help to illustrate both Buranelli's attitudes to social differences and some of the more striking socio-economic changes. Figures for the years 1780–84 offer a fairly articulate picture of Burano's and Venice's society. The population is listed in families and divided, according to *ancient regime* classification, into four social categories: 'noble', 'citizen', 'civil' and 'popular'. In the years 1780–84, the total number of families in Venice (today's *Centro Storico*) was 30,699. Only 3.1 percent of families were noble and 4.3 percent 'citizen', while the largest categories were 'civil' with 27.3 percent and 'popular' with 65.2 percent.

Table 5.1 Venetian families in 1784.

	Noble	Citizen	Civil	Popular	Total
Venice	954	1,333	8,399	20,013	30,699
Dogado	–	–	192	18,442	18,634
Burano	–	–	–	1,215	1,215

While Venice had a fairly large number of 'civil' families, 8,399, Burano's 1,215 families (5,396 individuals), as well as Torcello's 48 and Mazzorbo's 37 families, were all classified as 'popular'.[4]

Table 5.2 The population of Burano in 1784.

Population	0 – 14	14 – 60	60 and over	Total
Males	779	1,763	115	2,657
Females	–	–	–	2,739
				5,396

Another significant set of figures is that concerning various categories of 'regular religious persons': priests, clerics and nuns. In 1780–84 Burano had 29 priests, 11 clerics and no less than 79 nuns. Torcello had 3 priests and 72 nuns, while Mazzorbo had one priest and 71 nuns in the parish of San Michiel and one priest and 62 nuns the parish of San Pietro. The number is not as high as in Coronelli's report of 1696 (see Chapter 2), when Burano had over 100 nuns and 40 priests, but is significantly higher than for Venice, where nuns were 1,555 for 30,699 families.

According to statistics concerning trades or professions, Burano had 17 'practitioners in the Liberal Arts', 247 'artists and manufacturers, with their workers and apprentices', 68 shopkeepers and sellers of foodstuffs, but no agricultural workers or vegetable gardeners. In the list of 'professionals' are included several specialists concerned with different aspects of seamanship or boating: *squeraroli,* boat builders,[5] *peateri, burchieri,* boatmen employed in the transport of heavy goods, such as wood and building materials, and *aquaroli,* that is, carriers of fresh water from the terra firma. Under the general description of 'industrious persons' are included sailors, fishermen, servants and persons without income or trade, and the list ends with 'collectors of alms' – a rather high 5 percent of the population, which confirms much I have read and learnt from

Table 5.3 Trades of Burano's inhabitants in 1784.

Liberal Arts	17	1 %
Artisans and Manufacturers	247	13.8 %
Shopkeepers	68	3.8 %
Sailors	305	17 %
Fishermen	1,038	58 %
Servants	13	0.7 %
Unemployed	12	0.7 %
Collectors of Alms	90	5 %

informants about Burano's impoverishment from the end of the eighteenth century to the early decades of our own.

There are no figures for trades and professions for 1785–9. A page had evidently been prepared, but the census was never completed.

The census of 1981, when Burano had a population of 5,208: 2,578 women and 2,630 men (see Chapter 1), is naturally based on much altered categories. It is nonetheless of interest to examine some of the data which are indicative of deep changes in occupational and social structure.

Table 5.4 Burano's Employment Statistics in 1981.

	Male	Female	Total
Employed	1,411	360	1,771
Unemployed	74	32	106
Seeking first job	88	94	182
Housewife	0	1,394	1,394
Student	73	66	139
Retired	311	94	405
On Military Service	34	0	34
In other Condition	92	41	133
Total	2,083	2,081	4,164

According to a general division between 'active', which includes, 'employed', 'unemployed' and 'seeking a first job', and 'non-active' population, that is, 'housewife', 'student', 'retired' and 'on military service', Burano had an active population of 2,059 and a non-active population of 2,105. The distribution of men and women in the two categories, with 1,573 men and 426 women in the 'active' category, and 510 men and 1,595 women in the 'non- active category, shows a remarkable gender imbalance.

It is also striking that the number of women seeking a first occupation was considerably higher than that of men, with 19.3 percent of the total population of 'active' women, as against the men's 5.6 percent – a clear sign of the women's eagerness to enter the labour market, as well as of their difficulties in finding employment.

The highest number of workers were employed in the industrial sector, meaning, with few exceptions, in Murano's glass factories. By contrast, the number of fishermen had gone down considerably from 1,038, that is 58 percent of the male population in 1781, to 163 men and 3 women, that is 8.8 percent of the total population in 1981. About 20 percent of people worked in public administration – a description that offers little insight into social structure and hierarchy, as it includes a wide range of jobs, from 'white collar' to simple 'services'. A significant category was that of people working in trade, one in which numbers have much increased in the last twenty years, and in which men and women are less unequally distributed than in other occupations.

Table 5.5 Burano's occupational structure in 1981.

	Males	%	Females	%	Total	%
Agriculture, Fishing, Hunting	163	8.7	3	0.1	166	8.8
Mining and Industry	554	29.5	101	5.2	655	34.9
Gas, Water, Electricity	20	1.0	1	0.05	21	1.1
Construction	84	4.5	2	0.1	86	4.6
Transport and Communications	125	6.6	2	0.1	127	6.8
Banking, Insurance	17	0.9	9	0.5	26	1.4
Public Services & Administration	276	14.7	116	6.2	392	20.9
Trade	241	12.9	156	8.3	397	21.1
Other	5	0.3	2	0.1	7	0.4
Total	1,485	79.11	392	20.9	1,877	

The distribution of Burano's working population through the hierarchical structures of different trades and professions shows that the majority are simply classified as 'worker', *operaio comune*, with 483 men and 43 women, followed by 'skilled worker', with 390 men and 43 women, and 'self-employed' with 274 men and 34 women. The number of clerks is 155, 98 men and 57 women, but the two top positions, 'manager' and 'in managerial career' have, respectively, three and five persons; one manager works in trade and two, one of whom is a woman, in the public administration. Of the five people in managerial careers, one, the only woman, is in trade, one in the transport sector, one in banking, and two in the public administration. Persons described as *liberi professionisti* are altogether sixteen, all males; nine work in fishing, one in industry, two in business, three in administration and one in banking. The statistics, however, are by no means sufficiently detailed, for example, there is no category for 'lacemaker', since crafts are in general included under the heading 'mining and industry'.

Educational standards are somewhat lower than in other parts of Venice, especially in the historical centre.

Comparison of eighteenth-century census figures with contemporary statistics is near impossible because of the development of new forms of employment, such as 'banking and insurance', 'public administration' and 'industry', as well as radical changes in categorisation. For example, in the eighteenth century women were merely entered as 'female' and not in any way differentiated either by age-group, trade, employment or marital status.

Table 5.6 Educational qualifications in Burano, 1981.

	Males	%	Females	%	Total	%
University	18	0.4	1	0.02	19	0.4
Diploma	101	2.1	61	1.2	162	3.3
Lower Middle School	577	11.9	400	8.3	977	20.2
Elementary School	1,110	22.9	1,127	23.2	2,237	46.1
Can read and write	509	10.5	652	13.5	1,161	24
Illiterate	129	2.7	160	3.3	289	6
Other			1	0.02	1	0.02
Total	2,444	50.5	2,402	49.54	4,846	100.02

It is nonetheless worth noticing that, while in 1784 Burano had 68 shopkeepers, that is, 3.8 percent of its working male population (1,790), the number of 'persons in trade' in 1981 had risen to 397, of which 241 men and 156 women, or 21.15 percent of the island's 1,877 working population.

Other figures relevant to Buranelli's notions of social structure and status are those on housing. As we saw in Chapter One, great value is attached to the quality and state of repair of peoples' dwellings. At the time of my fieldwork home-ownership was not by itself a sure mark of status if the family did not have the means to carry out all the necessary improvements, and it was generally the case that owner-occupied dwellings were kept in better order and repair than rented ones. Differences in standards of living thus appeared with great immediacy in a comparison between the humid, decayed and poorly furnished interiors of some dwellings and the comfortable rooms of those that had been freshly restored.

Because the areas of Giudecca and Terranova had the poorest housing conditions, the highest number of rented and unrestored tiny two-roomed cottages and the highest incidence of house-sharing, they were sometimes described as 'low-status' neighbourhoods. Terranova was supposedly the 'New Land' on which the islanders settled after their earlier village, Buranello, disappeared under the sea. The minute size and the poverty of its cottages was thought to be due to their antiquity, and indeed it was sometimes alleged that its inhabitants have retained the highest numbers of archaic words and expressions.

Living in rented and overcrowded homes inevitably lowered a person's status, but compassion was usually thought more appropriate than disdain, although insults relating to the state and quality of housing were sometimes leveled during fights. Over the last fifteen or twenty years, a tendency for families to buy their homes whenever possible has led to some redistribution of people through different neighbourhoods. As in Venice, where previously depressed areas have been considerably gentrified, young couples bought homes wherever they became available, and therefore distinctions by neighbourhood were made very rarely, but the state of repair of individual homes was certainly an indication of the changing fortunes of Burano's families.

Idealised memories of the past as a time when, as some people maintained, Buranelli were all equal (possibly a compensation for extreme poverty), were usually modified or openly contradicted by acknowledgments that 'of course, there were *siori*, mostly the shopkeepers who would let people have food on credit. Sometimes they would charge interest, keeping their customers in a state of indebtedness, but, as people told me, they were *forced* to do that, since excessive generosity would have ruined them as well, and then who would have fed the poor islanders? Their function as mediators and providers was not viewed as an entirely negative or exploitative one.

In the 1980s, Buranelli were in fact quite affluent, and everyone paid cash for their food, so that shopkeepers did not have to extend credit to their customers any more. Thus, although differences may have become greater than in the past, the essential fact remained that everyone had risen above subsistence level. With memories of deprivation still vivid in many peoples' minds, and, although some types of job – especially clerical ones – were held in higher esteem than others, understandably, *any* employment by which a family could make a living was valued. Comparison of recent occupations with traditional trades show that, while fishermen were very much fewer, and numerous people have entered 'public' employment, some continuity remained nonetheless. For example, the numerous boatmen, or *barcaroli,* who, in the past, used to travel up rivers and carry goods in and out of Venice, have made a natural transition to jobs in the city's removal and transport firms.

Employment with the state or the civil service (*enti statali e para-statali*) which offered security, pensions and sick leave was considered most desirable. Thanks to their experience of seamanship, Buranelli were relatively numerous among the staff of ACNIL, now ACTV, that is, Venice's public motor-boat transport; and, although they objected when services were first introduced at the turn of the century, since they were regarded as unfair competition for those who made a living ferrying people on their rowing boats and large gondolas, Buranelli eventually became very attached to ACNIL, which they designate as simply *L'Azienda,* 'the Firm'. They enjoyed wearing its uniforms, and, when tourists were still regarded as interesting and worthwhile, they considered the Grand Canal or Venice-Lido routes highly prestigious. At the time of my fieldwork, about fifteen Buranelli worked as dustmen or street sweepers (an occupation which they may have inherited from inhabitants of now depopulated Poveglia).

Both trades, dustmen and employees of ACTV, require very early rising, but this was certainly not viewed as an inconvenience by people who had long been conditioned by the demands of fishing, especially because it left them time to do other jobs or wander out freely in their boats during the afternoon. Indeed, several women referred to the men's hours of duty' or indeed any time they were away from home' as *a pesca,* or 'on a fishing trip'.

In order to tease out various, and sometimes contradictory, strands in Buranelli's attitudes to stratification, a distinction has to be made between social relations with fellow villagers and those with outsiders, whether Venetians or inhabitants of other islands. While in the past Buranelli were in many ways dependent on the city for their economic survival, as they are today for any welfare, health and social services, they took great pride in the fact that they were not beholden to outside masters. Even though they relied on nearby agricultural areas for any food other than fish, they often affected a sense of superiority towards the residents of Mazzorbo, Sant'Erasmo and Tre Porti, who had traditionally farmed the fields of Venetian landowners as rentiers or share-croppers.

Buranelli were proud of the fact that, although the poorest did at times work for others, most families traditionally owned their boats and fishing nets: as they would say, 'He who is master of the sea is master of the land as well' (*Chi xe' paron del mar xe' paron de la tera*). As Marisa so well remembered, when, at times of hardship, groups of Buranelli used to go and help farmers in Tre Porti and Sant'Erasmo with the harvest, they would complement their meagre earnings by stealing fruit and poaching in the fish-farms, both of which they carried out with great dexterity. However, as they firmly stated, even when they did work under a master, they would only take on jobs they could enter on a contractual basis, and which they could freely abandon if they found that their employers or co-workers (especially working-class Venetians who always affected superiority) eroded their sense of independence and of equality (Muranesi too were strong rivals, as were the Chioggiotti.)

At the same time as they were known (or thought) to be defensive of their personal prestige and autonomy, however, Buranelli viewed Venetian society with respect for its hierarchies and with a great deal of admiration for its most powerful members. The language used to describe different statuses and positions at the time of my fieldwork was mainly derived from tradition, and was markedly different from that of the census takers. While the fundamental distinction was still between *poareti* and *siori*, people at the very top of society, those who command great political or financial power, were referred to as *i potenti*, the powerful. They were usually viewed as quasi-mythical heroes, outside of most peoples' experience. Among them were some of the most generous and extravagant buyers of lace, from Elisabeth I to Evita Peron. When such potentates visited the island, through a curious process of *de-doublement* (and perhaps, in the past, to overcome their own sense of insignificance), Buranelli would emphasise their common humanity; and if they approached such visitors with all manner of requests and petitions, they would always address them with dignity and measured respect.

Below the *potenti* came the *siori*, or *signori*, an expression that, in the Venetian context, connotes ideas of wealth and prestige, but also implies education, breeding and generosity, more than mere domination, as it might in areas where wealth was derived from land-ownership. Those who are *very* rich, or *sioroni,* may be inclined to vice, but others may be charitable and dedicated to good works. They are usually not thought of as exercising a profession, although they own factories, lands and villas in the country. After them came large numbers of people who worked for their living. Leaving aside professionals such as doctors, lawyers, teachers and various types of official and bureaucrat, whom Buranelli value very highly both for their skills and for their capacities to help in various ways, these are naturally viewed as no different from the island's inhabitants.

However, although a strongly egalitarian spirit generally characterised social intercourse and led many people to state that all work is equally worthy, distinctions were made between the new rich and a few families of long-standing status. For example, families of boat-builders, who, despite difficult times, managed to hand down their boatyard for generations extending back to the sixteenth or seventeenth centuries, were certainly viewed with respect, as were some of the more prosperous shopkeepers or wholesalers. They too had known bad times, but they managed to maintain a dignified status, and they never had to resort to public charity. As we have seen from Piero's opening words, hunger, as opposed to good nourishment, was a major trope in discussions of economic and status differences. People who had been very poor described their better-off neighbours as having been in a position to *eat* better than others, even at times of scarcity, or, at least, having been able to eat enough. They also described them as more reserved than others and inclined to seek the company of their equals. They naturally had well-appointed homes and impressive family tombs in the graveyard. They would wear better clothes, but would not indulge in too much display, and although they had money saved up, they were usually criticised for keeping it well hidden (*el morto soto'l leto*). However, their standing was never such as to create a strong and enduring hierarchy, particularly because the extensive kinship networks, and their keen awareness of kinship ties, would inevitably make their connections with less fortunate relatives obvious.

The drawing of boundaries and decisions as to how much and how far generosity should extend were in fact well-established and generally followed the law: it was imperative to help relations in the first degree and members of the household. Poor cousins and affines might receive favours – for example, childless aunts and uncles would often help to support their siblings' offspring – but knowledge that almost everyone in the island had poor relations contributed to a feeling that most families were basically equal and differences were precarious and contingent.

Inequalities were temporarily sharpened during the early years of fascism, when a few individuals with a penchant for domination sought opportunities to advance themselves. With the backing and support of external power brokers, they could validate their claims to political control at the local level, monopolise resources and take key positions in development enterprises that figured very prominently among the rewards exuberantly promised in the early years of the fascist regime. As some of my informants recalled, with their influence, then backed up by the state, fascist officials could decide to include or exclude people from jobs, grant commercial licenses, assistance, and so forth. Criteria for social acceptance then became stringent, and female sexuality was more rigidly controlled and more vulnerable to petty attacks than it had been in the past. A woman, at the time barely eighteen years old, was expelled from the Lace School when it was discovered that a non-related man lived in her house. He was a young orphan who had just been dismissed from an institution on reaching eighteen, and, as he had nowhere else to go, her mother had allowed him to make himself a bed under the beams in their attic – an arrangement that, according to the woman, in earlier years would not have drawn such active censure.

Some differences were thus exacerbated during fascism, when economic and cultural codes were manipulated and boundaries changed. However, a conjunction of moral and material criteria in evaluations of others was always basic to Burano's culture (cf. Pardo 1996). Thus, beyond the fundamental division between the poor and the rich, individuals and families were very often described as *de sesto* – a qualification that could be translated as *respectable,* but which contains more positive connotations than a mere absence of faults. *Sesto,* sixth, a metaphorical expression that derives from the recurrence of sixty-degree angles in navigation, boat-building, and architecture,[6] in this context means order, good measure, and grace. It is often applied to people who are impecunious and hard-working, but who can make ends meet and who can always cut a good figure. It is used also to describe gracefulness and a good social manner. Its diminutive, *sestin,* is used for children, especially girls, and may designate some endearing mannerism or personal trait.

By contrast, some people were said to be *poarèti* because they lacked the positive qualities of order, control and good measure, as well as the means to live well and to appear to do so. Occupational categories were not always relevant to social evaluation; indeed, people who were otherwise well off could easily have been viewed as *poarèti* if their married partners were incompetent and untidy, their children unruly and their appearance uncouth, while someone poor and unemployed might be described as completely *de sesto* and deserving of help. In sum, statements of social worth and superiority were usually based on a variety of criteria and on one-to-one comparisons between individuals or families, rather than labeling by class and economic position. Social interaction was egalitarian, and respect was dictated by distinctions of age and degree of familiarity rather than social distance.

Children were usually brought up to be polite, and most people were naturally gregarious and friendly. Neighbours always greeted one another, and when a person saw an acquaintance in the street, even if there was no time to stop and talk, he might call out his name, often imitating the typical intonations of boatmen, since acknowledgment was always a welcome sign of respect. Because cooking and food consumption took place in the nuclear family, a child's early social life usually began with visits to grandparents, aunts, uncles and cousins, and interaction was most informal with relations who lived in the neighbourhood. Except for invitations to important ceremonial occasions like weddings and funerals, which might include second and sometimes third cousins, distant relatives were not usually part of one another's social lives. People might say 'We are related, but we are not really familiar'. The expression used in this context is 'We have no domesticity' (*domesteghéza*) – a concept that defines the relationship by reference to the house rather than to kinship and was often used of non-relatives as well.

Since friendship beyond the family was highly valued, though sometimes feared as potentially labile, cousins bridged the gulf between kin and non-kin; in some instances they were lifelong friends, and those who lived in different areas were valued contacts and sources of information or privileged business partners. Although children were often warned against bad friends (and friendship in general was feared as undermining of family solidarity) they were actively encouraged to create bonds with schoolmates and neighbours. While many Buranelli told me that they did not especially enjoy school learning, they looked back on their school days with a sense of nostalgia and they retrospectively idealised the time (unfortunately too short) spent in education. Age-groups were quite cohesive, and even people who had gone to school for only three, five, or, at the most, eight years (i.e., middle school – see table) said that they felt a bond with their schoolmates.

Age-class solidarity and bonds of godparenthood were strongly encouraged by the Church as important means for developing and maintaining a spirit of *communitas*, and special masses were celebrated for all those who reached an age turning point, like thirty, forty, fifty and so forth. Women often maintained close ties with at least one – sometimes more – confidants, and friendships that started in early adolescence in some instances became lifelong attachments and permanent sources of affection and support. Lacemakers who lived some distance away from each other told me that they would cross the island every afternoon in order to sit and work by the side of their favourite friend and exchange knowledge and advice. Friendships between men were usually less intimate and exclusive, yet most men maintained that their relations were characterised by greater ease and camaraderie than were women's. Men, too, greatly valued companionship and sought the full acceptance and approval of their peer group, but closeness was not usually expressed by exchange of confidences; gossip among men was described as communication of news, and fellowship was heartily celebrated, when groups of men broke into song, laughed at one another's jokes and drank together late into the night.

There were in the island numerous associations and clubs, from political to sporting, cultural and charitable ones, and men were usually the most active participants. The political parties offered occasions for active interaction at their different and sometimes competitive *festas*, when people came to Burano for the day from different neighbouring areas. *Festas* also put into motion considerable networks of exchange, as large amounts of spaghetti, polenta, fried fish, Burano's typical biscuits and wine were consumed, while people danced in the piazza to the loud ballroom or rock music of some specialist band. Among the most important occasions are the September regattas, when, after participating in a complex cycle of competitive races against other *quartieri*, the island's oarsmen take their last and decisive stands in the Canale di Burano, and when all villagers are united against their rivals – often the champions of Pellestrina. Because Venetians, as well as the rowers of other islands, are involved in the races and city notables participate as umpires or spectators, hospitality is often extended in Burano's homes, and links between islanders and friends from outside become visible.

Friendships, both within the island and with residents of other areas, however, were thought to be as fragile as they were desirable. A permanent seal of amity is expected to take place when a friend is chosen to act as a baptismal, confirmation or marriage sponsor; in particular, by establishing coparenthood, bonds of friendship are sacralised and should become irreversible. Unfortunately, this was not always so, because even relations between *comàri* and *compàri* were sometimes marred by disagreements and became very distant.

Causes for strife were in fact ever present. The layout of houses and neighbourhoods, which made people's activities visible and any loud discussion or noise audible by neighbours, offered opportunities for continual and sustained observation – and at the same time led to a certain amount of secrecy, concealment and backbiting. Interaction with neighbours was therefore conducted with great caution, because everyone hoped that relations would be civil and cooperative, but competition, and a readiness to take offense, could destroy friendships and undermine good neighbourly relations. Sometimes fights arose because of encroachment. For example, although people have no legal right to monopolise public street space, such was the shortage of room indoors that small counters were erected opposite or beside front doors to store fishing gear, buckets and nets, while some women did their washing in large tubs just outside their kitchen. Complaints from neighbours would then lead to retaliation and to long-standing hostility.

One frequent cause of friction was that, while women liked to make a proud display of their freshly washed linen, which they used to hang across streets and *corti*, children might accidentally hit the laundry with a muddy ball. Reproaches, or even an occasional slap, would sometimes provoke endless recrimination from the victim's mother, and even necessitate calling the neighbourhood policeman. The worst enmities at the time of my fieldwork, however, were those caused by conflicting business interests, evictions, sexual jealousy and marital breakdowns.

The island's geographical position, and the fact that, like Venice, it has no land to cultivate, made it imperative that links with other areas should be carefully developed and maintained. Fruit and vegetables are brought in from Sant'Erasmo and Cavallino, while, before Burano was drawn into consumption of mass-produced packaged foods distributed on a national or international scale, meat, grain and wine used to be purchased in the fertile areas of San Dona', Treviso or Portogruaro. When the main source of income was fishing, the most important economic transactions took place at the Rialto market, where some of the most prosperous Buranelli had fish stalls, some being wholesalers, while others sold their catch at auction. Recently, the wholesale trade was moved to the island of Tronchetto, but Buranelli are still very prominent in the retail sale of fish. Business relations are always expeditious and brisk, but, by about eleven or midday, when most of the selling is done, the men gather in a small corner bar for a sandwich, a glass of wine and a coffee. They often stop to converse and years of familiarity inevitably lead to feelings of fellowship and mutual respect. In the same way, people who now hold jobs in Venice may become friendly with colleagues at work, but the two spheres, that of work in the city and that of home in Burano usually remain separate, and, with few exceptions, men said to me that they felt happier with their friends in the island.

Women, as we have seen, sometimes gathered to work in groups, and, especially for lacemakers, interaction was essential. A division of labour by which, traditionally, women were left on the island while men were away at work, naturally led to some stratification among the women that was quite independent of the positions of their male relatives. Particularly lace workers formed their own orders and ranks – indeed a hierarchical structuring may have come about partly in imitation of conventual institutions, which provided a ready model and cognitive picture.

Those lacemakers who were capable organisers would take on leading roles in coordinating the labour of small chains of workers, usually between five and seven, and, although members of the group were not equally close to all others, the sense of a common purpose drew them together and they were often seen to compare their products, ask one another's opinion, and boast about their abilities and their speed. Skill in their work and business acumen, as well as sexual and domestic virtue, then converged in making some women highly reputable and granting them the respect owed to natural leaders, although in the old days a kinship idiom was used also in teacher-pupil relations, and *maestre* were often addressed as 'aunt', *amia*.

Burano's women were criticised by more reserved and timid inhabitants of other islands for their *braùra* – a term for which, 'bravado' or 'audacity' only offers an approximate gloss, and which describes both their actual skill and their boasting – 'when they think they can do a thing well, they will go and tell it to all the four winds'. That may have been dictated simply by the need to sell their handicraft and to maximise their profits, and, although at the time of my fieldwork most of the lace produced was marketed by wholesalers, women still thought it expedient to maintain good connections outside, as well as within Burano, especially with some shop or rich family. Indeed, private customers are usually prestigious ones, since only they can afford handmade bridal veils and household items. Rather like *kula* objects[7], some of these were well known and a few senior craftswomen took pride in restoring them before they were used for the wedding of a daughter or granddaughter of the veil's original owner. While the lacemakers sometimes gave way to resentment through scurrilous jokes and invectives for having had to work long hours for some rich woman's wedding, they said that they nonetheless felt something of a bond with those families for whom their mothers and grandmothers had also worked. In general, then, women seemed more disposed than men to accept the realities of social hierarchy, and to manage unequal relationships with great skill and forbearance.

In the past, not only lacemakers, but also street-traders who used to go to the city to sell haberdashery or collect 'rags and bones', and even beggars, benefited by some acquaintance and sought to establish some reassuring contact and reference point in the city. For example, a woman, who, in earlier years, had taken to begging out of desperation, explained to me that her mother had shown her 'a few doors in Venice where she knew she would not be rejected'.

In recent years the nature of relationships between Buranelli and inhabitants of the historical centre has greatly changed, and it is the Venetian shopkeepers who have to compete to buy the little genuine lace that is still made in the island; old garments and rags are not traded any more, but are given away for recycling or charity, and most clothing is bought from department stores. Efforts to establish connections and to benefit from a network of support outside the island, however, continue, and are mainly directed towards those officials and professionals, such as doctors, school-teachers, lawyers, or judges, whose help may at times be essential. Such relationships certainly have features of *clientship*, but there is little Buranelli would give in exchange for protection other than hospitality, friendliness and perhaps a few votes for some political candidate. Above all, they would not compromise their autonomy.

As is frequently stated, 'friendship does not spoil business' (*amicissia no guasta mercanzia*). Enterprise, friendship and trade are thus inseparable, and business associations, exchange and even bureaucratic procedures sometimes generate enduring solidarities that may be described as purposeful, but which I should hesitate to define as solely instrumental.

In this section I have examined aspects of stratification in Burano and I have come to the conclusion that, given the islanders many cross-cutting ties, affiliations and activities, to construct a coherent picture of 'social class' is virtually impossible. At the time of my fieldwork, occupation was certainly not the primary criterion for stratification. The island, as we have seen, was almost untouched by industrial development, and labels such as 'proletariat' or 'working class', in the current Marxian understanding of those notions, were not really acknowledged as appropriate by most Buranelli. Yes, they said, they certainly worked very hard, but, somewhat like the Naples' *popolino* described by Italo Pardo (1996: 5, 15–16), Buranelli mostly regarded their work as free enterprise. In their understanding, *operaio* designates someone who works as a hired hand in anothers' business or fishing boat, but that was often a temporary, or part-time arrangement, and often alternated with some other employment, such as an autonomous enterprise or service in the public or tertiary sector.

Although many Buranelli described their work, whether in fishing or in Murano's glassworks, as heavy and tiring in the extreme, most of them said they abhorred work in Marghera's industrial factories, which they regarded as dangerous, unhealthy and subject to rigid discipline. It is of note that some of the late eighteenth-century categories I have described above still inform Buranelli's ordinary language and discourses on their work activities. For example, although, as we have seen, the number of fishermen is greatly diminished, the notion of fishing as an art, or highly specialised craft, is often present; following traditional usage, older men refer to their fishing nets and tools as *arti*, and for any well-executed technical job, young men will often be rewarded with the praise 'you are a real *artista*!

Given the fragmentation of social classes, the piecemeal nature of many working lives and the plurality of bases for belonging, there is an increasing tendency for social groupings to form on the basis of subjective orientations and lifestyles. Choices of dress, cultural interests and entertainments, and in particular holidays, are beginning to contribute to the creation of a new cultural patrimony. Indeed, for Burano, patterns of consumption certainly lead to new forms of social integration and solidarity in addition to those traditionally offered by the Church, the Associazione Cattolica, and, to a lesser extent, the various associations and political parties. At any rate, tastes, manners, moral reputation and a capacity to have enduring friendships both within and outside the island all contributed to persons' and families' standing in their community.

Notes

1. Most Buranelli's attitudes to social differences certainly did not tally with the Marxian view that belonging to a social class provides the basis of people's cognitive outlooks.
2. It was not unusual for informants' firm statement 'here we are all equal' to be rendered puzzling by their later remarks about neighbours' envy, competitiveness and self-regard.
3. Reluctance to admit that economic differences actually undermine the belief that people are all equal is also related to a Franciscan vision.
4. The census includes figures for the years 1785–89, which indicate that the number of families had slightly increased, while the main variation was a significant decrease in the number of noble families, from 954 to 894, and of well-to-do Jewish families, from 83 to 64. For the years 1785–89 the population of Burano had risen to 5.491, but the number of families had gone down to 1.149.
5. Venice's aqueduct was built in 1884. Till then water was drawn from wells or brought to the city mainly from the Seriola, a channel derived from the river Brenta.
6. Venice itself was divided into six city districts, *Sestieri*, while the sextant was used to measure the angular distance of stars from the horizon, and thus to estimate the position of ships. *Sesto* is also widely used in the Venetian dialect to designate the curvature of wood used in the construction of boats, or the simple tool by which it is measured, and the span of most arches is also

commonly referred to as *sesto*. The Italian equivalent of *sesto* is *garbo*; *sesto* has only remained in the expression *mettere in sesto,* to tidy, or in the word's technical usages.
7. The *Kula* is a ceremonial exchange of bracelets and necklaces described by Malinowsky for the Trobriand Islands (1922).

6
HONOUR AND SHAME IN MEDITERRANEAN ANTHROPOLOGY

A 'southerner': and what else? But the whole world is South – and especially in its worst aspects...

Leonardo Sciascia, *1912+1*

At the time of my fieldwork Buranelli's descriptions of life in their island were often based on contrasts between past and present – a rhetorical mode that expressed their way of coming to terms with the social change of the last thirty or forty years. In particular 'honour' and 'shame' were frequently linked with memories of past experiences, obligations and feelings about parental control, or impositions of order and discipline by authorities, mainly from outside their island. Notions of honour had greatly changed and the word itself was rarely used by the islanders, since it was felt to be pompous and inappropriate to their valued simplicity, while shame, and in particular sexual shame, was partly wished away as a painful legacy from an oppressive past. But would one be justified in saying that honour and shame had become quite irrelevant or absent from Buranelli's lives and sensitivies? If I had suggested outright that in Burano the values of honour and shame had entirely disappeared, that would probably have been taken as an offense. In view of such uncertainty – and to respond to much anthropological work on southern European societies – the question I posed at the time of my fieldwork was: 'did "honour" and "shame" as analytical constructs have any validity in 1980s Burano?'[1]

On a first reading, much anthropological work on Greece, Italy and Spain left me with a measure of incredulity and doubt about the importance of honour and shame as normative values; my response, in some ways like those of my informants, no doubt was partly that of a native, while inevitably my research was informed by an established tradition of English anthropological questioning and focused on themes that had been singled out as central to life in the Mediterranean. Before turning again to my account of fieldwork, therefore, I shall briefly discuss those aspects of 1950s and 1960s ethnographies that first drew my attention to 'honour and shame' and which, since the 1970s, have attracted much criticism from numerous anthropologists working in the European Mediterranean.

Some British Anthropological views of Mediterranean Honour

Important landmarks in studies of 'honour and shame' were the publication of Pitt-Rivers' *People of the Sierra* in 1954 and John Campbell's *Honour, Family and Patronage, a Study of Institutions and Moral Values in a Greek Mountain Community* in 1964. There followed in 1965 a collection of essays by several anthropologists in Peristiany's edited volume, *Honour and Shame: the Values of Mediterranean Society*. It was mainly after the publication of this volume that 'honour and shame' became a standard expression for describing numerous, and sometimes rather diverse, southern European and Islamic societies, and that a great deal of effort was expended in trying to identify common features which might justify such wide-scale generalisation and comparison.

Briefly, the values of honour and shame, which are thought to be normative, as they strongly influence behaviour, are shown to be sexually polarised, in ways comparable to those in which social roles, the division of labour, or various other rights and obligations are differentiated on the basis of gender. As a result, although it was generally stated that feelings of shame could be experienced by both women and men, honour was described as the prerogative of men, shame the concern of women.

Surprisingly, the word 'honour' hardly appears at all in *People of the Sierra* (Pitt-Rivers 1954;1971). By contrast, discussion of shame takes up almost an entire chapter – namely the chapter on women. The concept of shame, however, is not theoretically analysed, and its treatment is essentially descriptive. In general, as Pitt-Rivers emphasises, separation of male and female spheres in the Andalusian village where he conducted his fieldwork was very strict and clear cut; it affected upbringing, education and work, as well as playful and recreational activities. Women played a predominant part in the home and in neighbourhood relations, while the men were often working some distance away from home. As he writes:

> The male social personality has been related to the conception of manliness. The feminine counterpart of that conception, which expresses the essence of womanhood, is *verguenza,* or shame (1971: 112).

Men's social personality would thus have been based on notions (whether innate or culturally determined) of masculinity, while women's experiences would have been entirely conditioned and constrained by 'shame'.

Pitt-Rivers concedes that:

> the word has first of all a general sense not directly related to the feminine sex…It is a moral quality…which may be lost…It is closely connected with right and wrong…but is not synonymous with conscience. It is rather its overt or sociological counterpart.

As he concludes, 'shamelessness faces the world … wrong faces one's conscience (113) [2] – a distinction often found in works on early morals and usually due to a vision of 'primitives' as lacking individuality and moral autonomy. Indeed a shift of 'wrong' from 'shame', as a basis of Pueblo morality, to 'conscience', as typical of northern European sensibility, added to the distinction between external and internal sanctions, inevitably conjures up images of ill-developed 'primitive' morals. It is revealing that, as Pitt-Rivers notes, such distinctions were first made in the context of evolutionary theory in an essay by Marett on *The Beginnings of Morals and Culture. An Introduction to Social Anthropology* (1931: 408). Going back to the source, we see that, according to Marett, a truly rational morality, in which behaviour is ruled and restraint exercised by conscience, would have developed after an initial phase in which man was blindly driven by instinct, followed by a second, but still 'pre-ethical' phase, in which social order was based on the domination of habit and custom.

No anthropologist in the 1960s would have applied Marett's evolutionary scheme to any society. Yet Pitt-Rivers' description of shame in a Spanish context, *verguenza*, inevitably contains a hint that pueblo morals were in some ways 'pre-ethical'. One can ask if a contrast between southern European 'shame' and Northern European 'conscience' may not have been just a way of seeing the former as 'other', or, as some critics might say, of 'creating otherness' (Sciama 1974, unpublished seminar paper, de Pina-Cabral 1989).

In Pitt-Rivers' account *verguenza* is closely associated with the sexual conduct of women, but a man too can be disgraced through the misdemeanour of his mother, sister or wife. A woman's behaviour can then determine the moral standing of a family within the community; in particular, if a wife is unfaithful, this testifies to her husband's lack of manliness or his ineptitude as a bread winner, and blame then attaches to him. Hence, the widespread use of horn symbolism, which shows that a man has fallen under the domination of brute nature through his failure to defend a value vital to the social order. Male potency, itself a physical character, is thus seen as the means to subdue potentially disruptive sexuality. Due to the presence of a double standard, sexual activity is thought to enhance male prestige, although it may cause a woman to lose her *verguenza,* and thereby taint the reputation of her male relatives. Also masculinity, then, can be dangerous and can be the cause of social evils when it leads to sexually predatory attitudes and excessive competitiveness for individual glory and pleasure.

According to Pitt-Rivers, therefore, male and female are not diametrically opposed, but they are strongly differentiated in moral as well as material terms. For example, 'sexual activity brings prestige to a man, but, except where its purpose is that of having offspring, it may bring disgrace to women'. The idea of complementarity is clearly implied in the importance given to the institution of marriage, for it is within marriage that the sexuality of women is put to its proper purpose, procreation, while male potency is harnessed and contained. Marriage then allows both women and men to be in control of their selves and fully socialised into orderly adult existence. The values juxtaposed by Pitt-Rivers in his early work are, as we have seen, *verguenza*, 'shame', and *hombria*, 'manliness', or 'virility', but in his *People of the Sierra* (1954) 'honour' had not yet entered his discussion of Spanish life in any significant way. As Davis observed, 'it is not until he writes under Peristiany's editorship (1965) that his analysis of honour becomes fully developed' (1977: 94).

The first extended ethnography in which 'honour' and 'shame' are opposed is John Campbell's *Honour, Family and Patronage, a Study of Institutions and Moral Values in a Greek Mountain Community* (1964). In that book, which, as we have seen, preceded Peristiany's edited volume by a year, honour and shame, the difference and complementarity of feminine and masculine moral codes, as well as relations between honour as a dominant preoccupation of Mediterranean people, and conscience, as more typically a concern of Protestant northern Europeans, are some of the most compelling themes.

However, in Campbell's account of Sarakatsani shepherds, honour and shame are integrated into a system of moral and metaphysical conceptions far more fully analysed than those of peoples described in earlier ethnographies; as social values, they are the this-worldly correlatives of religious attitudes and beliefs.[3] Indeed, Sarakatsan morality is related to a cosmology that articulates itself through a coherent system of classification, in which an opposition, central to their symbolic and belief system, is that based on sexual differences, and expressed in an analogy with their familiar world of herds:

> For the Sarakatsani, sheep and goats, men and women, are important and related oppositions with a moral reference. Sheep are peculiarly God's animals, and their shepherds, made in His image, are essentially noble beings. Women, through the particular sensuality of their natures are inherently more likely to have relations with the Devil; and goats were originally the animals of the Devil which Christ captured and tamed for the service of man...[but] Sarakatsani will often say that although Christ tamed these animals the Devil still remains in them ... Women are not, of course, simply creatures of the Devil but the nature of their sexuality which continually threatens the honour of men, makes them, willingly or unwillingly, agents of his will.
>
> The intrinsic principles of honour refer to two sex-linked qualities that distinguish the ideal moral characters of men and women: these are the manliness (*andrismos*) of men and the sexual shame of women (*drope*). The quality required of women in relation to honour is shame, particularly sexual shame. (1964: 26)

Indeed, shame as a temperamental quality or disposition of character is thought to descend in the female line, and, according to Campbell, manliness and shame are complementary qualities in relation to honour, since 'women must have shame if the manliness of the men is not to be dishonoured' (271).

> The opposition of the two sexes is also reflected in a pattern of sex-linked attributes which are antithetical but complementary. Men are courageous, austere, possess manliness and pride, but lack patience. Women are fearful, lacking in resolution but have greater depths of love; they are sensitive to shame, modest and patient. However, the masculine values transcend this opposition and provide the ideal virtues of the total community. (276)

Sarakatsan gender classification, which is so starkly expressed through the symbolism of sheep and goats, God and the Devil, is thus examined as part of a coherent and well-developed folk psychology based on the clear-cut opposition of male and female.

Campbell's book was followed in 1965 by Peristiany's collection of essays, in which sexual polarisation of honour and shame is thought to hold true for all, or most, circum-Mediterranean countries, and it was at that point that both concepts were extended beyond reasonable capacity. For, despite the book's broad comparative basis, and despite evidence that honour and shame have widely differing ethnographic contents in different places, for different social classes, and at different times, it generated a tendency for wider generalisation than was justified. Too narrow a focus on honour, Allison Lever observed, 'has distracted our attention away from the values of non-dominant groups' (1986: 83–106). For example, Pitt-Rivers' attempt (1965) to construct a structural scheme of relations between concepts, behaviours and status variables, led him to formulate relations and oppositions that now appear quite forced, if ingenious, even from a purely linguistic and logical, as well as an empirical, point of view.

Attempts to reach an overall definition of honour thus brought to light a tangle of conceptual problems and unresolved questions: was the Mediterranean a homogeneous cultural area? Was honour 'egalitarian and moral', or was it 'morally neutral, hierarchical and materialistic'? Was it mainly based on others' approval or on inner needs and on sentiments about right and wrong?

In his book *People of the Mediterranean* (1977: 89–101) honour is defined by John Davis as one of three main determinants of social stratification[4] – so common a feature in Mediterranean societies as to support his view of 'the Mediterranean' as a homogeneous culture area. Trying to reach a firm definition of honour, Davis makes clear that his understanding of informants' constructs of honour is unequivocally materialistic. He points out that conceptions of honour that emerge from earlier ethnographies fluctuate between 'egalitarian' and 'moral' versus 'materialistic' and 'hierarchical' ones. In that way, he finds that earlier works, especially by Campbell, Pitt-Rivers and Lison Tolosana, are self-contradictory, in as far as 'honour is construed as virtue', but 'as social reality shows, it is in fact closer to rank and prestige'. For example, while Campbell emphasises the nature of honour as 'moral precept' and 'true nobility', it actually emerges from his work on Sarakatsani shepherds that it has a material basis, and is itself a system of stratification. Similarly, Davis observes, Pitt-Rivers asserts that there is an egalitarian ideal of honour, while 'everything in his ethnography shows that a hierarchical conception is at work.' (89 ff.)

On these points Davis's comments certainly touch on a genuine difficulty in social analyses: hierarchical and egalitarian values are inevitably present in most societies and either may prevail, while conceptions of honour may vary greatly under different social and political circumstances. Davis's observation about the disparity between honour as 'moral precept' and 'social reality', therefore, brings to light a discrepancy between ideologies and practices noted by anthropologists in the Mediterranean and elsewhere – and one that informants are often eager to point out. However, a disparity between social practices and egalitarian ideals does not in itself represent a contradiction in the *concept* of honour-as-virtue, or show that people do not hold such a concept or that they deem it irrelevant to their own lives.

Davis's summing up of the paradox, 'the same word is used for both the status of the powerful and the virtue of the weak' involves unquestioned assumptions that are not always borne out by comparative study and social observation: as my work in Burano shows, moral and political discourses are always integral to peoples' material circumstances and social position (cf. Hirschon 1984: 5; Blok 1975: 205–210). Indeed, status and virtue are not necessarily opposed or mutually exclusive, and 'the weak' are not always powerless – especially if we are disposed to take account of spiritual, religious or intellectual power and of the status these may confer.[5]

Some of the ambiguities and potential confusions of egalitarian versus hierarchical features of honour can be clarified thanks to a distinction introduced by Frank Stewart between horizontal and vertical honour; the first, associated with egalitarian societies or ideologies, is the right to the respect of others to which all are equally entitled – and which, in some instances, can be legally defended. On the other hand, vertical honour (Davis's 'status', 'prestige' and 'rank') is based on superior position and is associated with competitive hierarchical societies. As Stewart points out (1994: 59), claims to vertical honour depend on a person's right to the respect of her/his equals, therefore complete separation between horizontal and vertical honour is not possible – except, I should add, for the most cynical and disenchanted of social observers.[6]

In their discussions of honour, several of my informants showed full awareness of its contradictions and their comments were usually quite diverse and subjective: Anna related her sense of honour to her remote aristocratic origins, while for several men honour derived from the possession of skills, their competence in fishing and their capacities for generosity and fair dealing, or success in sporting competitions (Sciama 1996: 160; Lindisfarne 1994: 82). In general, however, while honour was understood to be inevitably bound up with emulation and achievement, it was always a hope that the moral worth of people might match and complement their material standing.

At the same time it was recognised that such hopes are often disappointed and most people, at the time of my fieldwork, were aware that the word was charged with potential ambiguity. Reports of Italian corruption and Mafia crimes showed that honour discourses are widely appropriated by violent groups, who conduct their activities according to rules and 'codes' which they subsume under their own idiosyncratic notion of 'honour'. Indeed, abuses both of the language and the reputation of honour rendered it open to much ironic comment. However, although denigration of others' high standing was sometimes prompted by envy, gossip about some peoples' allegedly 'false' honour clearly implied a desire that 'true' honour should be recognised. While they expressed strongly egalitarian attitudes, it was their opinion that honour should be based both on material prosperity, as evidence of a person's adequacy and enterprise, and moral achievement; honour-prestige and honour-virtue were not always regarded as mutually exclusive and it was clearly their expectation that moral goodness and social recognition should, at least partially, coincide. A need for a degree of harmony and a strong view of their village as a *civil* society, therefore, tempered any excessively competitive and disruptive behaviour. As in the Castillian society described by David Gilmore,

> a man's honour [but in Burano a woman's as well] ... relates to a substratum of social exchanges, traditions and ethics ... a rivalrous sex-linked honour of dominance exists, but is overshadowed by the pragmatic and convivial ethics of honesty ... and performance in all areas of social life ... especially cooperation rather than openly conflictive behaviour (1987: 99–100).

Indeed a kind of honour that has received little anthropological attention – although it is included in Pitt-Rivers' structural scheme (1965) – but often is the most relevant to people's lives, is that of *'honour-virtue'*. This is the least encumbered with connotations of rank and hierarchy; what is more, it concerns both women and men, from the highest placed to the most humble, and it may be close to the 'true egalitarian root' of honour itself. It is when honesty turns into honour, or 'honours', and when reputations become either the bases for establishing trust and gaining access to resources, or the motives for the community's rejection of humble and timid or inadequate individuals, that informants and anthropologists alike, may, with true pessimism, look upon honour as equivocal and open to question.

A growing sensitivity to gender issues since the 1970s led to radical criticism of Mediterranean anthropology. Indeed a rather facile and indiscriminate use of the 'honour/shame' label for areas otherwise different in economic, religious or social terms had led to descriptions of women as invariably powerless, economically dependent and subject to male authority. This was further aggravated by a tendency, widespread in the 1970s, to construct structural accounts of societies based on enchained oppositions; honour/shame was presented as homologous with brave/cowardly; strong/weak; public/private, and so forth (du Boulay 1974: 104).[7] However, although, as Herzfeld points out, oppositions are undoubtedly intrinsic to the cultures of the people we study, they are certainly not adequate to describe variation or to represent more nuanced social attitudes, practices or sentiments (1987: ix). Above all, however, given a tendency for oppositions to imply a hierarchical element by which one of the terms is a dominant or superior one, it was a foregone conclusion that women should be in all ways the dominated and subordinate category. As a result, sexual polarisation was not merely a question of difference, since it always involved a rigid attribution of positive and negative qualities to male and female respectively.[8]

Feminist anthropologists showed that shortcomings which characterised the whole of the research process, from the training stage to the fieldwork and its writing, were due to the fact that their work had been constrained within disciplinary styles and traditions almost entirely devised and laid down by men. In the most radical versions of their critiques, they advocated the need for a complete revision of research practices as well as anthropological theory. As Rayna Rapp Reiter writes:

> We need to be aware of the potential for a double male bias in anthropological accounts of other cultures: the bias we bring with us to our research, and the bias we receive if the society expresses male dominance... Our own academic training reflects, supports and extends the assumptions of male superiority to which our culture subscribes.
>
> If we only have a vague idea of what constitutes dominance, we cannot know if it reflects the experience of both men and women, or if it is instead something that the men assert and the women deny. In such a case is dominance a male fantasy? An anthropological fantasy? (1975: 13–15)

A male perspective had led to very poor observation of women's lives, so that many otherwise accurate ethnographies suffered from omissions and lacunae where women and gender were concerned (Sciama 1974). In particular, feminist readings of Mediterranean ethnographies at first produced almost a shocked reaction: the control and powerlessness of women at a purely practical level, and the recurrent honour/shame: male/female oppositions at a symbolic and moral level seemed to represent European misogyny in its most extreme and almost quintessential form. While male spheres were studied in great detail, women's realities were invariably viewed through the lens of male opinion.

A description of honour as a basis for power dependent on economic resources almost exclusively held by men clearly represents women as never being in a position to assert their personal honour. For example, when Davis (1973: 94–5) states that a rich man could seduce the wife of a poor man with no ill consequence, because a poor man would not be able to prevent such an act or take revenge, the implication is that women have no capacity for honour or independent will of their own. Indeed, the vision is one of women as tokens in relationships between men. Moreover, given that men's discourses on gender roles were reported as fundamentally similar and repetitive, criticism of the idea of 'the Mediterranean' as a uniform culture area was all the stronger when focused on the description of women – or lack thereof.

Readers were thus confronting accounts of gender relations which were shocking precisely because they were in some ways rather close to situations and discourses that they may have experienced or witnessed in their own lives, but which were far more explicitly formulated in discussion of 'other', but not so fundamentally different societies. Indignation and disbelief, as is implied in Reiter's passage, were then natural reactions. Could part of it have been in the eye of the observer? The supposition was that reports may have been onesided through lack of direct knowledge and understanding of the women – itself a consequence of the women's reserve and of their being protected from contact with a non-related man, or foreign woman, in the person of the visiting anthropologist. Hence, Reiter's firm statement: 'We need more studies that will focus on women.'

Her suggestion has certainly not gone unheeded and a change of focus has led to close and sensitive observation of women's spheres. Some writers have collected evidence which shows that sexual polarity was not always or everywhere as extreme, public and private spheres not as rigidly separate and unequal and land frequently owned by women to a much greater extent than had been implied in earlier ethnographies. Moreover, observation of the prominence of women in family ritual and of their key role in the household economy or craft production has led anthropologists to question the unqualified attribution of dominance and superiority to public male spheres.[9]

Recently, most anthropologists have guarded against an uncritical adoption of the honour/shame dichotomy either as homologous with gender or as a distinguishing feature of Mediterranean cultures; however, disbelief as to its accuracy for those works in which it was first formulated now seems unjustified, so that Campbell's work, and Pitt-Rivers' *People of the Sierra*, retain all their force, precisely because they describe pure examples of deep-rooted Mediterranean misogyny, and a closeness of practice, classification, and ideal codes, which is by now rather rare due to dynamic social change since the 1950s and 1960s.

Feminist critiques, thus, emphasise the practical difficulties of fieldwork in rapidly changing present-day southern Europe, and show the burdensome nature of rigid theoretical assumptions formulated by earlier anthropologists; but, although the 'honour/shame' dichotomy may have run its course in Mediterranean studies, it is nonetheless proving fundamental in other ethnographic areas, such as, for example, India, China, Japan and northern Europe. In my next section I shall therefore examine further some of the difficulties in early anthropological works in which 'honour and shame appear to have been discovered' (Charles Stewart, in print).

The Notion of Shame: History and Translation

Difficulties in the use of 'honour' and 'shame' as analytical categories and terms of an opposition, may partly be due to their linguistic history and the history of their translations. In the first place, honour and shame, whether in English or their Romance equivalents, are not simple opposites: honour always connotes some ideology; like beauty or truth, it may engender strong feelings, but it is not itself an emotion or experience. Furthermore, as Unni Wickan writes, shame, 'is experience-near', it *is* an emotion with well-known somatic expressions, which psychologists relate to disgust, anger and fear (Marsciani 1991: 35–50). [10]

As Herzfeld points out, relations between the two terms may vary, going from opposition to analogy, and almost identity. 'Honour and shame', he writes, are 'inefficient glosses on a wide variety of indigenous terminological systems'. It is not always correct to view them as the two terms of an opposition, given that shame (or Greek *dropi*) can be regarded as a positive virtue in men as well as women, and as a sense of restraint, or a 'brake', it can be equivalent to honour (*filotimo*) rather than its opposite. As a result 'massive generalisations of honour and shame have become counterproductive and their continued use elevates what began as a genuine convenience for the readers of ethnographic essays to the level of theoretical proposition' (1980: 339–351).

To clarify different meanings of the terms – and taking up Spitzer's observation that 'Etymological origins and changed nuances may be as relevant to our understanding as are comparative anthropological studies in one or more areas of the European Mediterranean' (1948: 1–14) – I have, therefore, briefly followed aspects of their linguistic history and, especially for the Italian equivalent of 'shame', *vergogna*, its transition from Latin to the vernacular (see also R. Needham's 1972; D. Parkin 1982, and R. Just 1998).[11]

In this instance, the development of attitudes that eventually were firmly rooted in Italian culture (or cultures) was parallel to the formation of neo-Latin languages in the middle ages, and long-term changes in language seem to have proceeded apace with the evolution of mores. In particular, 'honour' and 'shame' figure quite prominently in early Italian texts, so that even a few examples show that different meanings and nuances were already well-established by the twelfth and thirteenth centuries. Both terms would deserve a more extended treatment than is possible or appropriate here, but, partly to redress an imbalance in earlier ethnographies (see Rapp Reiter, above and Brandes 1987: 122–3) I have mainly concentrated on the history of 'shame': was it really, as anthropologists implied, a predominantly female experience? And, if so, what were its meaning and its history? To what extent were doubts raised by critics due to inaccurate translations, to ethnographers' gender-blindness, or both?

Does 'shame' as a key concept in Mediterranean studies diverge from its non-English equivalents to an even greater extent than 'honour'? Is it possible that impoverished literary translations may have led to an incomplete understanding of Latin and Romance concepts, which, crystallised in the common language, may eventually have been transposed from the translation of texts to the translation, or translations, of culture?[12] Unlike 'honour', a word derived from Latin *honor*, 'shame' derives from an Old English or Teutonic root. Its contemporary glosses are 'feeling of disgrace', 'state of disgrace', 'circumstance causing this' and 'modest feeling'. From the Old English roots of shame, *sc(e)amu*, there also developed *skand*, 'infamous man or woman', and 'disgrace' or 'scandal', as well as *sham*, which means 'false' or 'counterfeit. As the Oxford Dictionary of English Etymology suggests, Middle English *schame* may be related also to the verb *ga-hamon*, to dress, to cover oneself, and to *hemethe*, shirt, equivalent of the Gaulish *Camisia* and Old French *chemise* (1985).

Association of shame with 'dress' or 'shirt', which implies notions of modesty, are certainly familiar, as is the idea of uncovering falsehood and deceit.[13] 'Shame', however, has been used quite indiscriminately to translate several cognate, but by no means synonymous, Latin terms, like *pudor, verecundia*, and *modestia*. For example, in translations of Ovid's *Ars Amatoria*, possibly to satisfy the requirements of meter and rhyme, translators have often used 'shame' for *pudor*, but not for *verecundia* (Mozley 1939), while *pudor* is also rendered by weaker notions such as 'decency', 'propriety', and 'good manners'. On the whole, therefore, 'shame' and 'shameful' have acquired greater generality than the subtly different Latin words they have been used to translate.[14]

Both *pudor* and *verecundia* are verbal nouns formed respectively from *pudeo* and *vereor*. An interesting – and possibly revealing – feature of *vereor, veritus sum, vereri*, 'to feel shame', which also means to guard, to protect, is that it belongs to a group of verbs designated by grammarians as 'deponent'. Like other deponent, or semi-deponent verbs (a definition due to their loss of an active ending and acquisition of a passive one, e.g., *misereor* = I have pity; *morior* = I die; *obliviscor* = I forget, *vereor* denotes a personal experience or emotion.

In this instance, the verb form itself implies an idea of a sentiment midway between activity and passivity. It may also suggest that a shift from an active to a passive form may have been due to a change in meaning from *vereo*, 'I guard', 'I watch' to *vereor*, 'I am guarded', 'I am afraid', and 'I feel shame'. Indeed, the experience or behaviour trait described as *verecundia* is eminently born of interaction, but, at the same time, it implies a moment of reflexivity, or one in which others' reactions are anticipated and internalised. What is of interest, however, is that in its transition from Latin to Italian, the deponent verb actually becomes a reflexive one, *vergognarsi*, which both retains the Latin deponent's implication of subjectivity and emphasises the personal nature and the internalisation of the experience it designates (Benveniste 1966: 55–64).[15]

Thus, if we try to imagine a visual representation of *verecundia*, we may form a picture of someone looking at another person or persons, but not wanting to be seen or looked at, and having to protect her/himself by fending off other peoples' inquisitive or potentially censorious and accusing eyes. Indeed, like *modestia* and *pudore, verecundia* and *vergogna* have a rich iconography in the Italian painting tradition from the Primitives onwards. The reticent and *vergognoso* person lowers her/his eyes, or, as are many women saints and Madonnas, is characterised by a rather faraway glance, either turned upwards toward the Heavens and the Divinity or narrowly focused on the child. Although *vergogna* does not necessarily imply a lack of curiosity, a person's desire to protect her- or himself from the inquiries of others should really be stronger than the inclination to look at them. As girls in convent schools were strictly taught, eyes should be down-turned, and direct eye-contact avoided, because it is difficult to be possessed of *verecundia* and be inquisitive about others at the same time. *Verecundia-vergogna,* then, also implies silence, discretion and reserve, while the change of *vereor* from a deponent to a reflexive form clearly points to a growing interiorisation of feelings of self-consciousness and self-censure.

Also *Pudor*, which, as we have seen, is usually translated into English as shame, is related to a verb, *pudeo ,* glossed both as 'make ashamed' and 'be ashamed' or be 'restrained by shame'. It derives from the Indo-European root *pu,* to beget, or *pav,* to strike. Related verbs from the same root, *pu₂* are *puteo, putere* = to stink; *putesco* = to rot, to putrefy, and *puto, putare* = to clean, prune, set in order; regard, value, believe, think (*non putaram,* I would not have thought it).[16]

Most relevant from our point of view is the original association of the root *pu* with generation, as well as the fact that words derived from it designate ideas both of purity and its converse, pollution. For example, in contrast to the adjective *purus,* clean, pure, undefiled, chaste, we also have their opposites, *puter, putris,* rotten, decaying, putrid (Meillet 1921: 323ff.). An interesting feature of this root is, then, its polarity, that is, the fact that it generates words with completely opposite meanings. This, as appears from Franz Steiner's (1956) discussion of *sacer,* may be a feature of words or concepts charged with strong metaphysical contents; indeed it reflects deep-seated and unresolved affective and social contradictions (cf. Douglas 1966).

To conclude, in origin *pudor* was closely associated with generation and sexuality, while *verecundia* connoted feelings more generally focused on right behaviour and social interaction. Their Italian equivalents also seem to have merged, although a core difference remains, since *pudore* is closely associated with the genitalia or more generally the body, and *vergogna* is the more general and comprehensive term.

Unfortunately, little is known about Italy's early vernacular, as Latin continued to be the language of clerical writing and Italian literature flourished considerably later than its French and Provençal models (Migliorini and Griffith 1984: 41–2 and 56–70). Indeed, the earliest occurrences of 'shame', *onte* or *vergogne,* in neo-Latin writings are those found in the eleventh- and twelfth-century French and Provençal texts that were the forerunners of Italian literature. By the beginning of the tenth century, however, priests were encouraged to spread the word of God and take confession in 'the rustic idiom',[17] and it was at that period that there began to evolve moral orientations and sensitivities destined to acquire great permanence.

Developments in neo-Latin languages went hand in hand with an affirmation of new moral attitudes and cultural forms. Between the tenth and the thirteenth centuries there evolved a new, vernacular, and above all Christian, conscience that would continue to shape the moral and religious life of subsequent periods. Change was not occurring solely in the sphere of religion. The growth of cities, the elaboration of logic and learning, the quickening pace of trade and the formation of new social classes all contributed to greater value being placed on individuals and new emphasis on self-examination and self-awareness (Morris 1972: 10–11 and 64–71). At the dawn of the new millennium, legal systems, as well as monastic practices, began to be characterised by a new interest in motivation and intention, while a heightened concern with inner states also led to reflection on the emotions, especially love, shame and self-consciousness in general.

A document of great interest from our point of view is a legend in which, in accordance with the medieval allegorical tradition, *Vergogna* is the name of the main protagonist. The narrative, which originated in France in the eleventh or twelfth century, eventually spread through Italy, where street singers also sang it in verse. Written versions go back to the early fifteenth century, but a German folklorist reported that the story was still told in rural Tuscany in the nineteenth century, and that he had himself collected it from 'the people' (Toschi and Knust 1866: 398, cited in D'Ancona 1913: 77).[18]

The narrative reveals a profound concern with the problem of incest. Giving way to the devil's temptation, a widowed father seduces his virtuous young daughter. The child of their union, whom the father has decided to name Vergogna, is set afloat (like Moses) in a waterproof cot or box that will take him safely to Egypt, from where he will eventually return, and – unknowingly, like Oedipus – will bring about another incestuous union when he marries his mother. While this second incestuous union takes place in ignorance of a kinship tie, the first occurred with the father's full knowledge that he was committing a very grave sin. Repentance follows the act almost immediately; the father admits that he was 'the greater sinner', and he decides to expiate by strict penance, but says he would be loath to let his error be known to others.

A distinction between sin and dishonour is therefore fully acknowledged when, having decided to conceal his transgression, the man says to his daughter that he 'had rather [bear] the sin than the dishonour in the eyes of the world'. In the end, however, a need to expiate is stronger than his fear of peoples' censure, and he will die a pilgrim on his way to Jerusalem. As well as illustrating an early example of a distinction between guilt and dishonour, which some anthropologists, as, for example, Pitt-Rivers, implicitly linked with a contrast between primitive or archaic and contemporary morals, as well as between southern Catholic and northern Protestant Europe, the story also documents some of the difficulties in the Church's efforts to establish control over family morals and marriage rules.

As D'Ancona writes, the narrative seems to have spread throughout Europe and have risen to popularity at the time when the Church was struggling to assert its authority, amid a lively debate over the definition of prohibited degrees. In 1065 a Florentine lawyer had tried to establish that kinship degrees should not be reckoned according to ecclesiastical laws, but should be based on Justinian's' *Institutions*. These maintained that kinship of brothers and sisters was not, as in Church reckoning, first degree, but was second degree, and that of their offspring fourth degree. The grandchildren of brothers and sisters would be related in the sixth degree, and could therefore be permitted to marry by ecclesiastical dispensation. That doctrine, which had already been in circulation for a long time, was threatening to take root, and indeed was often followed in practice, since cousin marriages were greatly favoured (Wolfram 1987: 21–30 and see my section on kinship above). Only Gregory VII (ca. 1020–1085) eventually succeeded in re-establishing the Church's discipline (D'Ancona 17–19).[19]

In the legend of *Vergogna*, there is full recognition of the relevance of intentionality to retribution. As D'Ancona writes, the edifying intent of the tale is to show that no sin is too great for God's mercy, and Vergogna is forgiven because he had sinned unknowingly. What is of great interest is that shame seems to be strongest when family bonds are violated and there occurs a breaking of the incest taboo.

Another early occurrence of vergogna is in a verse dialogue or *contrasto* by a thirteenth-century poet, Giacomino Pugliese.[20] The poet complains that his lady has not lived up to her promise to be responsive and faithful; as if looking at himself through others' eyes, he laments that when he passes by the woman's house, she visibly hides, in a way that 'covers [him] with shame'. The picture we form is one at the same time stylised and extremely familiar; as the woman explains, she is so closely guarded (we presume by her father, husband, brother, or female relatives – 'may Christ confound them!') that she hardly dares to go near the threshold. She is caught in a dilemma; pressure from her family to protect her chastity and reputation obviously conflicts with her lover's insistent demand that she give in to his courtship.[21] In this case, *vergogna* is a sense of discomfiture due to personal failure to live up to an ideal, that of the successful lover, but at the same time it is the outcome of contrasting and incompatible codes, the pressures of courtship and fears of pollution and dishonour that are a known feature of Italian life.

A striking example of the gender polarisation of *vergogna* and *onore* is a poem by Guittone d'Arezzo (ca. 1230–1294) in which he vehemently chides the Florentines after their defeat in the battle of Montaperti. Flying from their enemies, and losing their ensigns, as he writes, they also lost their honour, their freedom and the favour of their lord. As their honour turns to shame, and given the obvious etymological link between Florence and flower, not only are the Florentines feminised, but they have become 'deflowered wretches'. As in many cultures, images of rape become powerful metonymies for military defeat, or indeed for all domination and abuse, while honour and shame appear to be most strongly opposed in the context of feuding and of male military and political struggles, in which shame is experienced collectively and publicly.[22]

A strongly misogynist spirit also characterises one of the earliest known occurrences of the reflexive verb *vergognarsi* in a work by a thirteenth-century poet, Jacopone Da Todi, *Laudi* (Battisti and Alessio 1957). The poem opens with an exhortation to women not to turn their glances on men, because their eyes, like those of the basilisk, cause souls to be lost.[23] Here, women, 'servants of the devil', are regarded with a fear and pessimism that clearly recall sentiments described by ethnographers; a woman's glance is in fact more dangerous than that of the basilisk, which only kills the body, while women can kill the soul, and lead to eternal damnation (Campbell 1964: 31–32, 277; Du Boulay 1974: 101; C. Stewart 1991: 176).

Shame, and more specifically sexual shame, remains a topic of interest and focus of debate throughout the thirteenth century. In a French allegorical poem, the *Roman de la Rose*, courtship of the Rose (the Rose being the symbol of the loved woman) is ironically described as a long battle between Venus and Chastity.[24] Shame, who helps Chastity in guarding the Rose, is the daughter of Reason and Misdeed, 'A creature foul and horrible to see, / Whose glance alone caused Shame to be conceived' (Dunn 1962: 59) at once conveys an impression of a troubling and contradictory sentiment – sexophobia.

In psychological terms, Shame therefore appears to be the result of a conflict between repulsion and desire, because, as well as a need to protect a person's bodily integrity, it represents withdrawal from physical disclosure and sexuality. At the same time, however, it clearly presents a dilemma – one that will be resolved only when the lover finally conquers the Rose. Moreover, although Shame is described as 'honest and wise', it is certainly not presented as a virtue, but rather as an obstacle to the enjoyment of life, and, above all, to procreation and continuity. In the second part of the poem by Jean de Meun, love and sexuality are analysed in the context of scholastic debate, and, in as far as the *Roman* is also social satire, the treatment of Shame, and, of course, Reason, expresses the writer's criticism of the excessive devotion, asceticism and hypocrisy of Mendicants and Beguines (see footnote 14, p. 183), as well as impatience with the sanctimonious affectations of courtly lovers.

Judging from the large number of manuscripts, between two and three hundred, some expensively illuminated, the *Roman* was very widely read, especially by the aristocracy. It was one of the earliest books to be printed (probably as early as 1482) and was read throughout the Renaissance and the seventeenth century. It apparently fell into oblivion in the eighteenth century, but was rediscovered in the nineteenth century.

A manuscript of an Italian abridgment and translation of the *Roman de la Rose*, first edited under the title *Il Fiore* (Castets 1881), was found in the last century in the medical library of the University of Montpellier (Marchiori 1983).[25] The poem follows its French source very closely, but, leaving out its learned and philosophical passages, it focuses mainly on the themes of love and sexual initiation. Here too, *vergogna* is coupled with *paura*, fear, but, after a final confrontation between Love and Chastity, both Shame and Fear abandon the battlefield so the lover can at last achieve his purpose (Marchiori: 363–410). With a keen feeling for essentials, the Italian translator includes his version of embryo formation:

> When I had ploughed, I sowed the seed I had brought.
> The Flower's seed had fallen too, and I so mingled them together
> That much good grass was born thereof.[26]

Both the man and the woman, as the text implies, are thought to contribute to the eventual fruit of their union – a theory which, after much scientific debate and philosophical reflection, was eventually to emerge as prevalent and become a basis for Italian kinship and juridical theory – and one that was undoubtedly of considerable consequence in the relative standing of the genders (see my chapter on kinship).

Notions of honour and shame are also central to all of Dante's work. In those passages in which he discusses shame philosophically, Dante, following Aristotle, defines it as one of the six passions of the human soul (the others are grace, zeal, mercy, envy and love). As in many descriptions of Mediterranean societies, shame here is viewed as a negative virtue: it is good in as far as it prevents grave transgressions, because 'where there is shame, there is fear of dishonour' which shows a kind of natural nobility. However, while it is appropriate in the young and in women, because of their natural proneness to error (a passage which certainly coincides with those of many an anthropological informant) 'shame is neither laudable nor becoming in the old and in learned men, since they must guard against those things which might lead them to shame' (*Convivio* IV (1304–1307) 1966: 186).

Attitudes to shame, or indeed an understanding of what the concept means and what the experience of shame entails, are by no means clear and unambiguous. Contradictory or vague glosses of the word seem to have characterised its history: as Italy's *Crusca* Academicians observed, although *vergogna* certainly is a brake against transgression, its main cause is nothing but the fear of censure, which can be seen as cowardliness (*Vocabolario* 1612).

To close this section, examination of *vergogna* in a number of early texts shows that by the middle of the thirteenth century its various meanings, as well as social and gender connotations, were well-developed and articulated . The texts I have examined show that notions of *vergogna* both as humiliation or embarrassment in the eyes of others, and as a sense of guilt derived from self-censure, were widespread in Italy at the turn of the first millennium: contrary to Pitt-Rivers' view, shame and guilt were in fact described as very close and interdependent.

An increased emphasis on inner experience, implied in the development of verbs designating feelings from Latin deponent to Italian reflexive forms, is parallel to a growing sense of individual responsibility, intentionality and conscience. This suggests that such tendencies would have developed at a period earlier than the Protestant movement, contrary to implications by earlier anthropologists, who seem to associate individual responsibility with Protestantism and to characterise southern European countries as more enmeshed in community, less conscious of personal responsibility and moral agency, and indeed somewhat morally backward.

A relevant feature of the texts I have examined, in particular *Vergogna* and the *Roman de la Rose*, is the mutual influence and interaction of secular and Church forces in matters of sexuality – a feature of society I shall explore further in my chapter on lacemaking in Burano. Coming back to the problem of translation, which prompted me to examine Italian and other neo-Latin terms, I have so far found that, although English *shame* does not appear essentially different from Italian or French equivalents, it does seem to be more general, less nuanced and probably less central to peoples' conscious everyday behaviour.

Uses of Vergogna in Contemporary Italian

Contemporary dictionary definitions show that *vergogna* has remained remarkably stable in its essential meanings since the twelfth century. A recent analytical gloss is:

> a painful and humiliating perturbation which shows that a person (literally, the 'soul') is aware of committing, having committed, or being about to commit any action which may bring dishonour, contempt or ridicule; a…sentiment experienced when an enterprise has been unsuccessful, or an error has been committed…It is the same sentiment which comes from a state of dejection, poverty and need. A feeling of shame (*pudore*), which causes a person to blush when anything offends modesty, honesty, or custom and good manners, and the blushing itself… As a psychological event or experience it occurs after some incident in which a person feels that her or his image or persona is damaged. (Battisti and Alessio 1950–1957)

The verb *vergognarsi*, to be ashamed, to feel shame, is reflexive. It also has a causative form, *svergognare*, to shame, to reveal another person's transgression or shortcoming. Other forms used most frequently are *vergognoso* = shameful, reticent, shy, and *svergognato* = shamed. *Vergogna* is often used in the imperative form, and it is then a reprimand or a sharp reminder, especially addressed to children, to feel ashamed of actions that are considered disgraceful, improper, or even just unconventional. As a repressive word-act it is one of the most frequently used informal sanctions.[27] Italian children are often taught that society can be quite punitive if any of its codes or customs are broken, and they are generally instructed to stay clear of situations that may bring censure. When it is used to condemn behaviour considered undesirable, the word carries a strong threat of disclosure and public disgrace.

The past participle *svergognato/a* – predictably, most times in the feminine – may describe a temporary, or quite ephemeral, loss of dignity and 'face', but when it is used as a verbal noun, it may connote permanent stigma, and a person so defined may be considered open to all manner of transgressions, since she or he is no longer possessed of the reticence that would act as a brake. It is in this sense that *svergognato* corresponds most fully to the Spanish *sin verguenza*, for it not only indicates that there has been an episode when shame was appropriate, but also that shamelessness can be an unconquerable character trait.

In Burano, *Vergogna* (and, to some extent, *pudore*) is frequently applied to women and related to loss of virginity outside marriage and consequent loss of respectability, but it is by no means only, or even primarily, related to sexual behaviour, because it usually follows on any shortcoming. As I found during fieldwork, social shame as a feeling of inadequacy in coping with unfamiliar and intimidating strangers, especially outside the island, is as common as is shame about personal and sexual matters. Moreover, although the threat of shame is used very liberally during the upbringing of children, as well as in battles for social pre-eminence over potential or actual rivals, to overcome it is also considered an important aspect of socialisation, because too keen a sensitivity to shame is not always considered a virtue.[28]

In contemporary Italian life, and particularly with respect to women, shame is by no means equated with purity and restraint, so that a capacity to put aside all shame can sometimes be considered an act of courage and personal affirmation. As for other moral traits, such as generosity against selfishness, or truthfulness against lying, young people must be taught to strike a happy medium, since most parents want to teach their children to be truthful and giving and sharing, but they soon realise that excessive openness and generosity, like excessive fear of authority, or fear of sex, would make them quite unfit for economic and social survival outside the family circle. For example, a capacity to lie is considered essential, as in some Greek communities (du Boulay 1974: 77–8 and 192–200, and Hirschon, personal communication). The extent to which shame is instilled, and then partially removed, varies at different stages of the educational process; it begins with toilet training, then continues throughout infancy, when shame is related to disobedience, aggressiveness or incompetence. However, as the child grows up and has to face the intimidating sphere of school, excessive sensitivity and fear of humiliation need to be corrected, and he or she must, at least partly, overcome feelings of shame in the interest of adjustment and performance.

For example, given the strong emphasis on oral testing and presentation in Italian schools, *vergogna* was often mentioned at parent-teacher meetings as a mitigating circumstance for poor performance. Some young mothers in Burano told me that they had to exercise a great deal of pressure on their children to reverse the ill effects of shame and respect of strangers too firmly instilled at an earlier stage in their upbringing. Eventually, between the ages of about nine and fifteen, children are gradually taught to deal with shopkeepers, teachers, bureaucrats and all kinds of unrelated people, and to remain self-possessed in the face of some difficulty.

Commenting on the wickedness of the neighbourhood butcher, a young woman conveyed to me a true sense of pride when she succeeded in persuading her twelve-year old daughter to return some unsatisfactory meat. 'It is no good being ashamed, my love, you've got to learn to get your money's worth, or shame will get you nowhere!' As the girl, flustered, but happy to have obtained a fresh supply of chicken breasts, rejoined the small group of assembled women, her mother explained to me that the issue was not only the meat, but she considered that having overcome the girl's resistance was a worthwhile educational achievement.

It is generally recognised that a sense of shame, instilled in both male and female children at a very early stage, may lead to excessive reticence and timidity. Indeed, deprecating the narrow strictness of his upbringing, one of my male informants lamented, 'in the end it was so strong, I had to be ashamed of being ashamed'. To give me a strong example of the way things were changing, Giulia, a busy, socially committed woman nearing forty and a mother of two, explained how much she disapproved of her contemporaries' domesticity and house-pride. As she said to me, at the same time giving her elderly mother a challenging look,

you'd think their honour was fully attached to their brooms and their dusters and needles and knitting pins! You'll never catch me sitting by my doorway mending socks or fussing over a piece of lace and leaving my door wide open to show everyone how clean and tidy my kitchen is. Oh, no, you are not going to see me scrubbing and polishing just to look good, or apologising and feeling ashamed if the floor is not shiny and resplendent! Mind you, it took three years psychoanalysis.

'A fine job, your psychoanalyst did!' said her mother ironically, 'Now I have to come and tidy up after you to keep your husband and family happy.'

Not all men, however, considered that a woman's sense of honour should be so intimately tied with domestic virtue; one of my most lively informants, a successful and prosperous fisherman, thought it would have been nice to sometimes go out with his wife on weekends or evenings: that, he thought would have been a departure from a 'men-in-the-pub'/'women-at-home' tradition, but his wife simply refused alleging some urgent domestic duty. 'The more you give them, new kitchen furniture, ornaments and domestic appliances, not to speak of clothes, the more they have to clean!'

But, while it was generally considered advisable for young unmarried women to show their competence and skill, as that would also have been proof of sexual probity and self-control, too intense a sensitivity to shame was regarded as a burden and a hindrance. An important area in which shame has to be gradually removed, if it had been too firmly planted, is that of female sexuality. Indeed, traditional courtship, or seduction, involves dissolving the inhibitory force of shame – a complex rite of passage described in the *Roman de la Rose* – and one that some Italian men seem to consider an honourable privilege, while others, with changing mores, now maintain that they find rather unnecessary.

In some situations, and especially where social rather than sexual shame is involved, also the teachings of the Church aim to modify the possible damage that may be brought about by an excessive fear of disgrace. Humiliation in the eyes of men, when not related to religious offences, is then negatively described as *rispetto umano,* 'human respect', and a strong distinction is made between that sentiment and respect for God and His teaching. Shame due to concern for people's opinions may gravely inhibit right and purposeful behaviour. As religious leaders and priests traditionally taught, shame, both as a brake and as a consequence of sin, should derive not only from a fear of human censure, but also from awareness that God knows and sees all, and that is, I believe, what makes it such a powerful and enduring norm and deterrent.

With few exceptions (Burman 1979; Goddard 1996 and Holmes 1989), honour and shame have been studied mostly in the context of the family and of extended kin groups. However, a concern with honour and the control of sexuality has been a key factor in arrangements concerning women's work outside the domestic sphere, and has posed severe limitations and constraints on their efforts to enter the labour market. In Burano, where in the past lacemaking was the main, or only means for women to earn a little money – sometimes badly needed – ideas about female purity, mostly upheld by the Church and by a compassionate, if somewhat narrow-minded and authoritarian aristocracy and bourgeoisie, were traditionally the guiding principles in the organisation of that work.

Very briefly, to answer my initial question on the validity of 'honour and shame' in present-day Burano, as it emerged from my informants' interest in and often searching conversations on the subject, both notions bear considerable relevance to their lives. 'Honour', although sometimes considered too high-sounding, and in some contexts even alien, a term, is nonetheless scanned out through more nuanced explanatory concepts such as pride, intelligence, skill, vigour, conviviality, and, above all, a capacity to stand up to others as equals – a very important self-affirmation, especially when faced with more powerful others – while punctiliousness, obstinacy, arrogance and self-regard are viewed as honour's negative sides (cf. Herzfeld 1987: 64).

A concern with shame is often present in women's reflections on their human condition and particularly in discourses about sexuality, indeed, many of them have expressed their need to free themselves of its hold over their lives and personal development. As will emerge in my next chapter, the past bears strong relevance to women's attitudes and notions of self, and that is why understanding of the underlying ethos and ideology of lacemaking has great explanatory power, as well as historical value.

Finally, as Burano's evidence shows, both 'honour' and 'shame' demand continual revision in light of changing social realities and mores.

Notes

1. As Johannes Fabian observed, 'Culture traits and cycles, patterns and configurations, national character and evolutionary stages, but also 'classical monographs' compel us to attach our arguments to Kwakiutl, Trobriands, Nuer or Ndembu – and, I should add, Sarakatzan shepherds and inhabitants of Spanish pueblos. They are so many topoi, anchorings in real or mental space of anthropological discourse' (1983: 111).

2. Recent writings follow a similar distinction. In Harré and Lamb's *Encyclopaedic Dictionary of Psychology*(1983) guilt is defined as 'the condition attributed to a person (including oneself) upon some moral or legal transgression. Shame is occasioned by an object which threatens to expose a discrepancy between what a person is and what he or she ideally would like to be…Shame may be considered the more fundamental of the two emotions. In western cultures, however, guilt has been the more highly valued because of its implications for autonomous action.' In addition, quoting a psychoanalytical view, 'Evidence suggests that men may be more prone than women to experience guilt, and conversely with respect to shame' (Lewis 1971).

3. Emphasis on religious and symbolic aspects of morals also characterises work by John Campbell's pupils, Juliet du Boulay, René Hirschon, Juao de Pina Cabral and Charles Stewart.

4. In Davis's view, others determinants of stratification are class and bureaucracy (1977: 76).

5. To see honour as a *material determinant* of social structure is really to lose its more diffuse but richer significance.

6. Other distinctions noted by Frank Stewart in various parts of his essay are: 'inner'/'outer honour'; 'personal integrity'/'reputation'.

7. As Herzfeld writes, 'despite their alluring neatness, structuralist techniques were in practice the expressive paraphernalia of a symbolism that we shared with the people we studied. They were not so much misguided…as grimly embedded in the objectivism that allowed Us to study Them (1987: ix). The fallacy of anthropologists [he writes] was to treat their analytical concepts as free of context (cf. Sciama 1981: 105–9).

8. Whether honour was described as egalitarian and an aspect of virtue, or hierarchical, based on material resources and in turn a basis for status and power shows the ethnographer's view of the morality of the people studied.

9. Clearly the honour/shame complex was formulated in fieldwork situations very different from those of most present-day southern European areas.

10. Given a distinction between shame that may follow transgressions and shame as a brake, by a linguistic slippage shame seems to have been identified with virtue, in particular the virtue of women, imposed through control.

11. I make no distinction between Italian and Venetian, as the word *vergogna* is the same throughout Italy.

12. The texts I have examined show that, except in the most misogynist contexts, honour and shame were by no means associated respectively with men and women.

13. See the recently coined media phrase 'name and shame', something like 'drumming the scandal'.

14. In its Italian usage *modestia*, from Latin *modus*, measure, is more directly related to humility, or deportment than are either *pudor* or *verecundia*. When English 'decorum' and 'propriety' are used to translate *Verecundia*, social aspects become the most prominent.

15. The Italian equivalents of some deponent or semi-deponent verbs are reflexive, or have reflexive as well as active and passive forms. For example, *audeo*, I dare, in Italian *oso*, is sometimes, more emphatically, *mi oso; taedet*, it is boring, *mi annoio*). To illustrate the point, it is useful to consider a verb that has an active and a passive, as well as a reflexive form, *spaventare, essere spaventato, spaventarsi*. This does not imply any knowledge of the extent to which fear may be due to circumstances outside the person, or to the person's proneness to be frightened due to nervousness and sensitivity. As with *vergognarsi*, we may be left in some doubt on the extent to which the experiences they describe are due to factors outside the person, or to his/her

subjective feelings. Linguistic evidence of an increased use of reflexive forms, thus, shows a growing internalisation of the sentiments described.

16. Some of the nouns derived from the root pu are, *puer* = child; *puella* = girl; *pupa* = girl, doll, puppet; *putus, pusus* = boy, and *puta* = girl. *Reputo*, from which 'reputation', has a meaning similar to *puto*, reckon, calculate, impute, and ascribe (to someone) good or bad actions.

17. A text that chronicles the coronation of Berengar I, in 915, relates that during the ceremony the people of Rome mingled songs in their native tongue with Latin and Greek compositions (Migliorini and Griffith 1984: 42).

18. French versions are known as *The Legend of Gregory Pope*. I have used the Italian edition by A. D'Ancona (1869)

19. As D'Ancona writes, the legend could be entitled *Gregorio Papa*, in his view the protagonist was Gregory the First, or the Great (c. 540–604), probably the first monk to rise to the papal dignity. It may have been attached to Gregory VII (1020–1085) at a later stage. In some versions Vergogna himself becomes a Pope, when he is called from his hermit's hideout – another detail that relates to discussion of whether monks, as distinct from priests, could be elected to the papacy (Morris 1991)

20. Giacomino Pugliese was classified by Dante as belonging to the Sicilian School of Frederick II, but it is not known if he ever was at the Emperor's court, given that Dante's classification may have been based on poetic style, rather than biographical fact. Nothing is known of the poet's life, but he is considered one of the foremost poets of the Sicilian School during the early and mid-thirteenth century.

21. The stanza in which *vergogna* first appears is:
Lady, does it not grieve you / to make deception or villany:/ When you see me go by / sighing in the street / You make a show of hiding / everybody chides you: / To you comes disesteem / Shame weighs on me / [my] love. (Dionisotti and Grayson 1949: 103–105)

22. Herzfeld sees strongly oppositional constructs as mainly generated by dominant elements in society – a view that Guittone's poem on the Florentine's military defeat supports (1987: 95–122)

23. You say you paint yourself that it may please your lord / but you deceive yourselves; he does not fall in love with it / if you but look at some simpleton, in his heart your lord suspects that you may be plotting against his honour. / He then feels wounded and laments, and guards you jealously / he wants to know which company and which people you frequent. (Dionisotti & Grayson 1949: 131)

24. Guillaume de Lorris began writing the *Roman* by in 1237, but the work was left incomplete. Forty years later, in 1277, Jean de Meun wrote a longer sequel. As one of its translators writes, 'the book stands today like some great French cathedral, conceived by its first architect in early Gothic…then extended on a grandiose plan'. The *Roman* is one of the few medieval texts that did not fall into total oblivion, but exercised considerable influence throughout Europe, and was translated into modern French in the Renaissance. (C.N. Dunn 1962: xii)

25. Arguing from internal evidence, some critics maintain that the name of its author, Durante, may have stood for Dante, and that the *Fiore* was one of Dante's early works. That, however, remains an unresolved question.

26. Sí ch'io allor il Fior tutto sfogliai e la semenza ch'i avea portata / quand'ebbi arato si la seminai. / La semenza del Fior v'era cascata: /amendue insieme si le mescolai,/ che molta di buon'erba n' é po' nata (414–5, CCXXX. See English trans. 462).

27. Offenders were publicly shamed by the Catholic Inquisition. An incident occurred in Burano when a woman was tied up in front of the church on a Sunday morning so that she could be shamed before all the villagers (ASV; Santo Uffizio 1627).

28. The continued appeal of the *Roman de la Rose* may be partly due to its radicalising influence and its emphasis on nature and authenticity. Difficulties encountered by Italian women during courtship, often due to tensions between the couple's families, are also caused by the pressure of contradictory impulses and social codes operating at the same time (S. Silverman, in Reiter 1975: 312).

7
BURANO'S LACE-MAKING:
AN HONOURABLE CRAFT

Where have those damsels come from whom I saw weaving cloths of silk and gold brocade in that yard? They are doing work that I find very pleasing; but I'm very displeased that their bodies and faces are thin, pale and pitiful...We shall never get out of here. We shall always go on working with our silk and will never be any better dressed for it...for none of us will get from her handiwork more than fourpence of every pound earned.

Chretien de Troyes, or *Yvain, the Knight with the Lion*

Arriving in Burano on a warm day, one is struck with the way the whole village seems to be decked in linens, richly adorned with embroidery and lace work. A continuous row of stalls sides the route from the boat to the main street and square. Several houses have been opened onto the street, their doorposts and windows turned into crowded displays; tourists are encouraged to enter and visit the rooms that were once kitchens and parlours, their walls lined with framed lacecraft madonnas, butterflies, flowers, boats and figures of lacemakers, fishermen or ladies and knights in eighteenth-century costume. Lace tablecloths, white, cream, light pink or aqua, hang airily from the ceiling, and a variety of handkerchiefs, shawls and ornate table covers are tastefully displayed in baskets and boxes. Behind the stalls and in the shops, women of all ages attend to the customers with remarkable patience, illustrating the craft and extolling the workmanship and the quality of materials. They show great forbearance towards uncertain purchasers and invite them to return, while other women, especially older ones, carry on working on some unfinished item.

Their ease and aplomb as saleswomen derives from the knowledge that *burano,* a name which distinguishes their lace from other varieties, has been well-known and valued throughout Europe since the late fifteenth century. Indeed, while, as we have seen, Burano has generally been overlooked by historians, some insight into its past can be gained through references to its lace and descriptions of lace-makers by travelers, journalists or art historians, especially from the late eighteenth and nineteenth centuries. However, although the craft has long been a basic factor in Buranelli's identity and far-flung reputation, as well as an important part of its economic history, never before the 1980s was such a profusion of lace displayed and sold in the island's streets.

In the next pages, I shall first explain very briefly how lace is made, then describe its recent developments in Burano, report some of the ways in which the craftswomen remember the past and look back on the ideologies and discourses that informed the organisation of work in earlier times. To end I shall show how changed attitudes to lacemaking and its virtual abandonment by younger women reflect aspects of social change and in particular views about sexuality, authority and female employment.

The Making

Most of the lace traditionally made in Burano is described as *punto in aria* , or 'stitch in the air', a name due to the fact that, unlike embroidery, the finished work does not have a linen foundation. Both the tools and the materials needed, a needle, a 'pillow', a wooden cylinder, a cheap cotton cloth and very fine cotton thread, are of the uttermost simplicity.

A design is first drawn on the paper, it is then picked onto the cloth by frequent stitches, which follow the outlines of the drawing, the needle going in and out of the same hole so the base will be easily detached and both the paper and the cotton lining removed from the finished lace. The lace itself is based on a sequence of buttonhole stitches, called, in the singular, *sacolà* , a word that bears a close resemblance to *sagolà*, 'fisherman's loop'. This forms the part of the textile known as *guipure* (a word probably derived from Bruges). The connecting tissue that joins different parts of the lace, usually figurative elements like flowers, leafs or stylised figures, is made of *sbari* , 'bars' or *rete*, 'net'.

Bars are mainly associated with lace made in Venice and are compared by the women to the bridges that join the city's islands, while in Burano's lace the background is a fine net; it too is called *burano*, as it is said to reproduce the technique by which fishermen used to make their nets. At a later stage of the work, called *rilevo*, 'relief', a thick thread, which gives the lace its striking three-dimensional quality, is applied at the edges and the outlines of designs, then the paper and cloth are carefully removed from under the lace with a blade. One essential fact is that *burano* is entirely different from bobbin lace, because it is made with just a needle; that, plus the fact that it is 'created' by its makers, since the cloth on which it is elaborated serves merely as a temporary support, is considered symbolically meaningful by the women.

Technical features of lacemaking, as of other textiles, are often associated with enduring symbolisms, while their history provides an interesting parallel and commentary to the history of social change and gender relations. Furthermore, evidence from Burano shows a strong association of European lace and lace making with sexual purity – an aspect that confirms the relevance of an honour/shame approach, but at the same time qualifies it by emphasising the way shame, as sexual restraint or repression, is anchored in Church teaching – an aspect sometimes overlooked by those anthropologists who analysed honour as mainly a secular code. As well as illustrating gender relations, the history of lacemaking concerns relations between social and economic classes, and for Burano, those of Venice and its island periphery.

Parts of the material I have gathered derive from archaeological or historical investigation, which, I trust, can be presented as 'positive' knowledge. Other parts, especially based on nineteenth-century writings on the origins of lace, are illustrations of conjectural diffusionist historiography, in which research was itself guided by symbolic connections rather than established sequences. In this instance, as we shall see, suggestions of scriptural origins for lace (which would be as difficult to prove as entirely to rule out) are nonetheless of great interest, in as far as they show the writers' inclination to 'read' the history of lacemaking in terms of religious values as originally a sacred textile that gradually became secularised, without entirely losing its association with ritual and with rites of passage. Readings of the history of lacemaking in a religious key as mainly practiced and promoted in conventual institutions also provide a contrast to alternative folk explanations of the 'invention' of lace as a native craft that owed its origins to the island women's ingenuity and industriousness.

Present Conditions

Despite Buranelli's great pride in their lace, and in apparent contradiction to its abundance in the island's shops and street stalls, at the time of my fieldwork many people feared that the craft might soon be altogether abandoned and forgotten. Since the 1960s women had been increasingly attracted to a variety of employments outside the island, for example, as clerks or packers in Murano's glass factories, nurses in Venice's hospitals, or caretakers and cleaners at the Lido beaches – all occupations which, in the past, would have been looked at askance, but which, once accepted, had contributed significantly to Burano's improved economy.

As a result, full-time lace making had largely been abandoned; twenty years earlier, as older women lamented, there were seventy or eighty *maestre*, teachers, between twenty and thirty embroiderers, and seven or eight hundred workers; in the 1980s, the number of *maestre* was reduced to nine and there was only one embroiderer, while ordinary workers were two or three hundred at the most. For as long as the women could remember, lacemaking had revolved round the Lace

School, which had traditionally been directed and run by nuns (and of which I shall write more extensively below) but due to social change and to the women's new needs and expectations, the School had gradually reduced its activities and had eventually gone out of existence in the 1970s. All its assets and archives had been left to a Consortium, eventually transformed into a Moral Foundation, with a view to initiating some new activity. However, although the School, as well as some of the entrepreneurs who followed its example in organising the women's work, were remembered as exploitative and authoritarian, those lace-makers who continued in their activity recognised that it had played an important part in their lives and they admitted that its closure had left them in a state of disorientation, because they found it difficult to produce and sell their handicraft without the support and leadership of an institution that had traditionally kept them in a state of dependence.

While most younger women had abandoned the craft altogether, the most dedicated lacemakers, who continued to work informally, complained that pay was still very low, and that their position as workers was insecure. Indeed, attitudes to lacemaking were at best ambivalent and contradictory: on the one hand, Burano's women truly looked back in anger at their own and their mothers' long suffering, above all the 'sacrifice', a word that came into their conversation very often, and summed up the physical frustration due to the work's sedentary nature, the suppressed tension, and the eye strain they had to endure from early childhood. On the other hand, they assured me that they loved their craft, which they extolled almost extravagantly, saying that, although in the past their lives were humble in the extreme, being able to make something delicate and precious filled them with pride, indeed, an awareness of being particularly able and refined assured them that they were an élite among workers, and it was an important part of their positive self-image and identity. All the same, a simple sense of reality naturally induced many of them to take up other employment.

In their accounts of their working lives, several lacemakers emphasised common themes and turning points. For example, Marietta's career was not unlike those of other women born between 1910 and 1950.

I was born during the War, in 1944, and I attended elementary school between 1950 and 1961. Those were very difficult years, my parents were respectable, but we were poor, and we children had to help keep the family going in every way we could. I first learnt to make lace from my mother and grandmother at home. Then I went to the School at twelve, in 1956. Because I acquired great skill very quickly, at seventeen I was allowed to go to join the more experienced lacemakers in the 'big school', and I began to make a few *lire*. When the Lace School closed down in 1972, I was 28 years old; I had two children and I still had to help support my parents, so I was forced to find another job. But I never forgot my old *passion* for lacemaking, and, as soon as that was possible, I took it up again, despite all its disadvantages. Most of my friends, on the contrary, lost interest in the craft; they found that other work was easier and better paid. Working in the glass factories, just tying up parcels, or dusting a few glass chandeliers, commuting home with their friends, and getting sick leave, maternity leave, health insurance, even a Christmas bonus, was child's play by comparison.

Although working conditions were much improved for lacemakers as well, discussion of the craft was almost inevitably accompanied by resentful and self-pitying comments about its low rates of payment, and about the lack of sensitivity of most people who simply did not understand how much patience and skill it required. That was particularly in evidence when plans to reopen the School, and partly turn it into a museum, caused the women more than ever to reflect upon their past. The idea that lacemaking might again be taught and practiced within an institutional setting had come about after a successful exhibition of lace, first held in Milan, then in Venice's Palazzo Grassi in 1977.[1] Due to a general awakening of interest in the history of textiles, some of the city councilors proposed that the Venice too, like Milan, should have its museum of lace, given that it had had a very prominent position in lace production. A number of Buranelli, who had observed that some of the most precious exhibits had been made in their island – 'with the sweat and tears of [their] mothers and grandmothers' – then demanded that the Museum should be in Burano. Previously, examples of lace had been granted a somewhat marginal position in exhibitions of glass, metal work, jewelry and printing, all mainly men's crafts actively promoted by the city and the provincial authorities. Now, thanks to a long overdue recognition of the exquisite artistry of Burano lace, the island would have its own exhibition rooms in the School building. At the same time, feeling that lacemaking should not be entirely relegated to the past, it was decided that also the School should be reopened, at least partially, and it was hoped that the lacemakers would eventually form a cooperative and sell their product directly to the public.

Surprisingly, however, a planned revival of lacecraft was by no means unreservedly welcomed by everyone in the island. The old building, a pleasant Gothic palace, which had been part of Burano's Medieval Town Hall, and was already undergoing restoration, would include a classroom where courses would be held on the making and the history of lace, while two large halls would be used for a permanent exhibition and provide spaces for temporary shows. However, while politicians were working to ensure the necessary funds, numerous laceworkers, shopkeepers, and even neutral observers expressed very strong doubts that the School would bring the workers any advantage at all.

Most of the women remembered the School as they had known it in the 1930s and 1940s when discipline was excessively strict, the nuns were punitive, and the pay was miserly. As one of them recalled,

> If they [the nuns] received two thousand lire for a piece of work, they gave you one thousand…what is even worse, one had to beg for admission; sometimes even to pay to get in. We had to pray, meditate, and take punishment. If a girl arrived late, or if she

was caught talking to her neighbour, her work would be taken away…They wanted to make so many saints out of us!

Moreover, while old women generally looked back upon their apprenticeship days with a degree of detachment and humour, some of their younger colleagues were still angry at the way the School closed in the 1970s, leaving them unemployed and without any compensation. As one of them observed,

> It always works out all right for the nuns, doesn't it? When they get into trouble, they get themselves called away by their order and they move on to a different house. A new mother superior comes along, and, would you believe it? She doesn't know a thing about it. That is how they got away with it, you know, when the School closed in the 1970s: they cheated the Government of their taxes and they left us workers without as much as a penny insurance or pension. Some of us had worked for ten, some for twenty, and some for forty whole years.

The mention of a new School, and, worse still, of a cooperative, which would attract the revenue's attention, frequently provoked negative or skeptical comments. In as far as I was suspected to have been in the pay of some unspecified government agency, I was either regarded with some reserve, or was made the recipient of women's bitter memories about the past and of their apprehensions about their working future.

Views on this latest revival of the School were expressed with dramatic clarity on the day of its official opening. A small party of notables who had arrived on the commune's speedy water taxis, as important visitors are often wont to do, so that they may quickly return to the city as soon as they have performed their official duty, gathered in the School's freshly restored premises. They included personalities as diverse as Venice's Councilor for Culture, the President of the Regional Union of Craftsmen and a number of other officials, as well as count Girolamo Marcello, a descendant of Countess Andriana Marcello who had had a prominent part in organising Burano's laceworkers in the nineteenth century.

To open the ceremony, the President of the *Consorzio Merletti*, formed after the closure of the School to take charge of its remaining properties and archive, welcomed everyone present, and, in particular, those who had come specially from Venice or Mestre. Then Venice's political representative (perhaps rather disingenuously) emphasised that the city council had welcomed with enthusiasm the request that the lace Museum should be in Burano. He spoke of the historical significance of its exhibits, and the great cultural value of the School archives. To end, count Marcello delivered a short inaugural speech. The Foundation named after his ancestress, and of which he had been made a trustee, was 'heir not only to all the old School's assets, but also to the ideas carried forward by his family for over a century.' His hope was that its patrimony could be used in a concrete and positive manner, beyond that of a merely conservative discourse' and that the School could become a place which the women would feel was their own – a place where they could work and attend courses, but also one where they could meet their friends, if they so wished, and then, 'why not?', even sell their handiwork.[2]

Count Marcello's talk was as brief as it was gracious. He recognised that the authoritarian and paternalistic attitudes of his antecedents in their dealings with the laceworkers may have been quite excessive, and he wished publicly to acknowledge that times had changed and to renounce any exercise of authority. Therefore, he mainly wished to bring his good wishes and express his affection for the island and personal attachment towards some of the older women present, whom he had known since he was a young child. Indeed, he was giving to Burano's Museum various items from a precious lace collection which had long been in his family.

Although, as other important people, the count appeared to be held in some reverence, on that occasion his visit to the island went quite unremarked, except by the few individuals who were involved in the School's organisation and by a small number of old loyal lacemakers. Very few women had attended the opening of the School – a fact vaguely explained by the local organisers as due to their being too busy with housework. The speeches, therefore, were delivered to a half empty hall, while the owner of one of Burano's large lace shops could be seen from the window pacing up and down the Piazza, as if undecided whether he should participate in the occasion or boycott it altogether. When the speeches were over, and generous refreshments served, people started to wander in, and their numbers increased as the Venetian dignitaries gradually began to leave.

The absence of several women I knew was a little surprising and could only be interpreted as a sign of reserve and suspicion. While all the speakers had agreed that the School should not become merely a token of the past – a hope everyone would readily have shared – it was, all the same, very difficult for Buranelli to be sure that this latest version of the old institution was going to be to their best advantage. Of its three proposed aspects, the Museum, the School and the Cooperative, the first was undoubtedly the least controversial, and, although reflections on their difficult past were never far from the minds of the craftswomen, some of them were happy to see that objects in the making of which they had, or might have, participated were now considered worthy of an elegant exhibition. However, the mere mention of a School usually evoked bitter memories, while the proposal that lacemakers should form a cooperative was most alarming, both to the shopkeepers and to those lacemakers who conducted their work informally.

The idea that a cooperative would now take important commissions and produce small items to sell on the souvenir market was viewed as potentially leading to unfair competition. Above all the proposal that only lace made by cooperative members should bear a mark of authenticity would certainly have overshadowed the traditional image of Burano as an island where lacemaking was a widespread but informal craft. Some of the workers, then, did not really like to be seen at the School, which is in one of the island's most visible spots, the Piazza, opposite the church and close to the most splendid lace shops.

Moreover, from the point of view of the craftswomen, the main objection to joining a cooperative was that they would have had to pay tax, and their earnings would have been so low that the craft might have disappeared altogether. Rather than share in a collective tax burden, most of them preferred to sell their handicraft privately to local traders whom they knew well, so that both parties found informal arrangements congenial, and solidarity was based on their mutual understanding and common advantages.

The pay, as one of the women assured me, was never commensurate to the skill and the time required, but, she added in resignation, 'I don't want to complain', 'I like to work freely as much as and when I please.' Indeed, when, after the Second World War, Venetian exporters had scaled down or abandoned the lace trade, their place was skillfully filled by a few newly prosperous local families. The way in which their wealth was made was not so much by selling authentic Burano lace, but by importing and marketing large quantities of relatively cheap Far Eastern products. A few fines from the Venetian Tourist Board for failing to state the true origin of their merchandise were certainly not enough to stop such trade; as with other offenses which Buranelli regard as minor transgressions or even small triumphs, in so far as they consider them as redressing old wrongs, the local police usually turned a blind eye. Moreover, with mass tourism, most of the items of lace sold are small pieces, and profit is made mainly on the basis of quantity rather than quality.

Attendance at the newly constituted School was thus very poor (although the whole enterprise cost the Venetian municipality and the Provincial Government a great deal of money) but, as time has shown, the mere mention in the tourist literature of a permanent exhibition in Burano has increased the number of day tourists, and also served to advertise the cheaper imported items for those visitors who wished to buy some souvenir, without the high expense of the genuine local product. While some of the lacemakers regarded the great influx of Far-Eastern lace as creating unfair competition, most of them soon realised that selling lace was more remunerative and far less taxing than making it. The most dynamic dealers, or those who had been able to put together some capital, then rapidly monopolised the trade and the long-standing association of lace with Burano continues to work as a powerful advertisement.

At the time of my fieldwork, therefore, most women – and without any expectation that I would believe them – claimed that they were making lace solely for themselves and their families. But, while sometimes their statements were transparently designed to conceal the fact that they worked unofficially, in other instances, they were quite true: pleasure in being at last in a position to keep for themselves a product, the value of which had been traditionally sanctioned by no less than royalty, but which they had never been in a position to own, has now made lace very desirable as ornament, and its possession a symbol of economic success. As a result, delicately fashioned framed lace Madonnas, flowers and butterflies, are displayed in almost every home I visited. Some of the craftswomen, however, could not resist the temptation of a direct sale, so that, when fears were allayed as to my possible role as a tax spy, many of them secretly offered to sell me some of their best pieces, in case I too wished to start a collection.

The History

Speculation on the origins and diffusion of needle lace has been as plentiful as it has been inconclusive. Because of deterioration, early examples are very few, nor can any firm evidence be based on the nature of the materials used, because lace has been made since antiquity with all kinds of fibre, such as linen, cotton or silk, hair, aloe and rafia. According to some of my informants lace in the 'old days' was also made with yarn spun from nettles. Findings of fragments among the clothing of mummies in Egyptian tombs suggest that lace must have been widespread in Egypt since the third and fourth centuries BCE, and may have subsequently spread throughout the Mediterranean and Armenia. In the Middle Ages, Coptic and Middle Eastern textiles were also plentiful in Europe and their ornamental motifs, mainly based on geometrical patterns like the designs of Moresque tapestries, leather work and ceramics, were widely reproduced in painting and sculpture. Evidence of Arab influence is also found in the Italian names of some types of lace, for example, *trine* and *macrame*, and in the fact that the earliest centres of lacemaking were those areas that held the most intense relations with the Arab world, namely Sicily, Venice and Genoa (1977: 22–24).

Lacking any reliable evidence to support conjectures that lace may have diffused from Phrygia, the Balkans, Greece, Byzantium, or even ancient China, Historians of Italian lacecraft mainly work on later periods for which archaelogical and textual documents are relatively plentiful. Renaissance pattern books, dedicatd to royal and aristocratic ladies, suggest that lacemaking, like embroidery, was initially an aristocratic pastime, while reference to Penelope and Lucretia, both archetypes of womanly chastity and family honour, emphasise an association of textile work with purity and virtue. However, a question sometimes raised by my informants in Burano (see below) was that of the craft's *social* origins: was it really, as the literature implies, originally an aristocratic occupation, or was it developed first, and possibly in much remoter times, by the poor inhabitants of Venice's islands?

An association of lacecraft with a strongly gendered notion of honour and virtue is present throughout its history. For example, it is of interest to read suggestions in an excellent late nineteenth-century book (Pasqualigo 1887)[3] that lacemaking may have been inspired by scriptural passages, mainly from Exodus, the Gospels, and the Gospels' Apocrypha.[4] As the Jews are about to end their Egyptian captivity, Moses instructs them to ask the Egyptians for 'jewels of silver and jewels of gold, and raiment' (*Exodus* III:16). With those, as well as all manner of rare and precious materials, '...fine linen, oil...spices...sweet incense', they will build the Ark, or Tabernacle, a movable temple, in which the Ten Commandments will be carried through the desert. The Ark, as God ordered Moses, would also be covered with a top of pure gold and with ten curtains of 'fine twined linen,' each edged with fifty loops, while a veil made 'with blue, purple and scarlet linen' would separate 'the holy place from the most holy' (*Exodus* XXVI: 36). Splendid garments would also be made for Aaron and his sons, whom God had elected to be the founders of his priestly caste. Such textiles were not exactly lace, nor were they white; what is of significance here is that they were nonetheless *sacred*.

Another biblical antecedent thought to anticipate lacemaking is that of the tassels of Jewish prayer shawls, that is, knotted fringes, made to remind the worshippers of God's commandment that they respect the Sabbath, 'and not ...go about after [their] own heart and...eyes, after which [they] go astray (*Numbers* XV: 32–41).[5] Here too, the idea of a textile entirely constructed on knots is associated with memory and devotion. Indeed, the essential notion in both Exodus and Numbers is the distinction between sacred and profane textiles: the first are highly precious and are therefore to be removed from contexts of profane (and possibly corrupt) aristocratic luxury.[6]

A close and exclusive association of such textiles with the Tabernacle, and then with the altars of early Christian churches, can certainly be seen as a source of inspiration for the organisers of lacework in medieval convents. Indeed, the Church was one of the earliest consumers of lace both for altar covering and for priests' vestments. As some Buranelli recalled, in the Middle Ages the patriarch of Aquileia demanded that part of the women's products be given to his church as tribute. Even more directly relevant, and comparable to the discourses that were recurrent in the organisation of Venetian lacemaking, was God's injunction that temple textiles should be made by those 'that are filled with the spirit of wisdom', and that materials and dyes should be given to the makers in a spirit of trust, which again emphasises the ceremonial nature of such altar covers and vestments, and the fact of their being removed from secular exchange cycles, set apart from all other things and only used in the sphere of the sacred.

A suggestive feature of the passages in *Exodus* is the order that a veil should separate the holiest part of the temple from its outer area, relatively less charged with the divine presence. In some ritual contexts, also persons must be shielded from view. For example, according to Jewish custom, as in Greek, Islamic and Hindu marriage rituals, brides are veiled and can only be seen by their bridegroom after the wedding ceremony is completed.[7] However, parallelism between sacred and human contexts of veiling reached a greater degree of symbolic elaboration in the Christian Middle Ages, when the metaphorical equation of the Tabernacle and the human body as a temporary abode for the soul became quite a commonplace of devotional literature.

According to rabbinical evidence, women appear to have been employed in the making of Temple furnishings since the days of the Babylonian exile (from 586 to

537 BCE). Ideas related to the making of decorative textiles and its association with religious devotion, therefore, seem to have had remarkable continuity, and it is of interest that they should have been so sedulously researched by historians throughout the nineteenth century. However, a question generally left unanswered is, when and how was a tradition of synagogue embroidery or lacemaking, and their association with female chastity transmitted to the early Church, and how did it enter the conventual tradition?[8] On this point I would agree with Pasqualigo's suggestion that strong textual evidence of textile work as an expression of holiness and virginity is to be found in several narratives of the life of Mary in the Gospel Apocrypha. Indeed, their influence and popular appeal, naturally due to the importance of the cult of the Virgin – may partly explain how a close association of textile crafts with female sanctity may have had remarkable continuity and may have spread from the Middle East to the Byzantine and Christian world.

It may be worthwhile to look briefly at the narratives. As is related in the Book of James,[9] Mary's parents Ioachim and Anna have promised that, if God granted them a child, they would vow her/him to the Temple. Mary is taken there with other young maidens when she is only three years old, but when she reaches the age of twelve, the priests are concerned that she may soon begin to menstruate and thereby become polluting. Joseph is therefore chosen to take charge of her. At the same time, or very soon after, as a council of priests have decided that the Temple should have a new veil curtain, officers are ordered to call 'some pure virgins of the tribe of David'.

> And the priests called to mind the child Mary, that she was of the tribe of David and was undefiled before God: and the officers went and fetched her, and they brought them to the Temple of the Lord, and the priest said: cast me lots, which of you shall weave the gold and the undefiled [the white] and the fine linen and the silk and the hyacinthine, and the scarlet, and the truly purple. And the lot of the true purple and the scarlet fell unto Mary, and she took them and went unto the house... Mary took the scarlet and began to spin it. (Book of James X.1)

While Joseph is away at work, Mary spends her time weaving. The angel of the Annunciation appears to her the first time, as she goes out to fetch some water. 'Filled with trembling she went to her house and set down the pitcher, and she took the purple and sat down upon her seat and drew out the thread'. The angel then again stood before her, saying:

> fear not, Mary, for thou hast found grace before the Lord of all things and thou shalt conceive of his word...And she made the purple and the scarlet and brought them unto the priest. And the priest blessed her and said: 'Mary the Lord God hath magnified thy name' (XI.2; XII.1).

Mary then rejoices and goes to see Elisabeth; she knocks at the door, and 'Elisabeth, when she heard it, cast down the scarlet [e.g.,'the wool'] and ran to the door'.[10] Fine textile work thus appears to have been widespread among virtuous women.

Continuity between the Bible's 'sacred stuffs' and 'fine linens' and Italian Renaissance lace, however, mainly concerns ideas rather than elements of design or techniques. The knotted fringes of Jewish prayer shawls connote ideas such as memory, regret and expiation of sins, striving for goodness, and above all the need to keep both the heart's desires and the eyes firmly in check – ideas with which the norms of the honour/shame code has already made us familiar, and which are also related to lacemaking. Indeed, Burano lace, as the women often reminded me, is entirely built as an elaborate sequence of tiny knots; but, while in general knots are thought to symbolically represent continuity, memory and fate, their symbolism is sometimes ambivalent and charged with contradictory meanings. On the one hand, in the Biblical and Christian traditions knots symbolise marriage as a lasting alliance – a chaste one in as far as it demands abstention from other sexual involvements, while three knots on monastic cinctures represent the three vows of poverty, chastity and obedience, and the girdle, as an attribute of Mary, signifies virginity.[11] On the other hand, knots can be instruments of witchcraft, as they can be made in a spirit of ill-will to magically bind and obstruct the movements of one's enemies and rivals, while their loosening brings about freedom and release (Lienhardt 1961: 282; Di Nola 1983; C. Stewart 199: 40–41 and 231).[12]

Such magical aspects of knotting are sometimes just hinted at in folk versions of the origins of lace, but they are almost never mentioned in its official or 'School' lore. Indeed, for as long as the lacemakers worked under the supervision of nuns, their ideal model was based on the notion that they should be devout, modest and wise – at least, this was the image constantly kept alive by their teachers and employers. However, to offer an alternative to most 'official' versions of its beginnings, some of my informants would place the origins of lace in an 'even more ancient' mythical past, but when legendary versions were reported, they were usually introduced by some cautionary disclaimer, such as 'I don't really believe it', or 'some old women say'.

Lacemaking from the Sixteenth to the Eighteenth Century

While attempts to identify the origins of lace and retrace its paths of diffusion are usually defeated by a lack of ancient artifacts, for the sixteenth and later centuries the designs and uses of lace are amply documented in countless portraits of those who wore it, usually members of the Church or the aristocracy. Its abundance shows that a large number of women must have been employed in lace making, but little is known about their working lives. Indeed, their predicament is effectively summed up in a historian's definition of lacemaking in Venice as 'this celebrated unofficial labour…which concerned a large part of the female population' (Gambier 1981: 21). Unlike weaving and embroidery, which were mainly practiced by men in workshops and supported by guilds, lacemaking remained within the domestic sphere or in charitable institutions, and the workers never enjoyed the protection of corporation or guild. Such failure on the part of the government to acknowledge the women's work is also the cause of the virtual disappearance of Venice's lacemakers from historical records – an absence that is all the more complete for Burano.

Conditions in the island periphery must have been considerably different from those of the centre, but lacking direct evidence, historians have based accounts of lacemaking in Burano on descriptions of its organisation and underlying ideology in the city (Molfino 1977: 27; Savio and Savio 1977: 39; Gambier 1981: 21). References to lacemaking schools in the palaces of two Doge's wives, first in the fifteenth century, then at the end of the sixteenth century,[13] as well as evidence from Renaissance pattern books, have led historians to conclude that after its beginnings as a gentlewoman's pastime, lacemaking would have percolated down from palaces to workshops, then to humble homes, and informal working groups would have become widespread throughout the city. At the same time, it was taken up as a suitable skill to be taught in convents and charitable institutes for the protection of orphaned, abandoned, and very poor women and girls, or repentant sinners. [14]

During the initial period the craft was practiced informally and there were no organised workshops, so the nuns 'worked here and there, or sat in their cells' (Gambier 1981: 22), and only in the seventeenth century did they start to work in groups, while 'one of them read a book of devotion or they all together sang prayers and psalms so they could pass the time with greater spiritual comfort' (Archivio Patriarcale, Monasteri-Visite 1452–1730, cited in Gambier, 33). As lacemaking became one of the main sources of income, especially in the poorer convents and institutes for the prevention of sin or the recovery of fallen women, it was managed with increasing strictness. Although such institutes were nominally secular foundations, organised by lay governors and governesses, the workers were usually supervised by nuns, and a religious inspiration was inseparable from notions of female honour, and from a view of poverty which, as is often observed in Mediterranean ethnographies, was viewed as a potential cause of sin and shame. As is stated in the sixteenth-century Founders' Acts of *Zitelle*, an institute for poor or orphaned girls:

> Virtue had at all times many fierce enemies, but never was it as threatened as when it was a companion to poverty and to beauty…Our pious plan was to found, with charities, a place suitable to welcome those girls between the ages of twelve and eighteen, who, endowed with beauty, but lacking the means for an honest subsistence, either through the loss or through the wickedness of their parents and relatives, were on the brink of losing themselves forever (*Notatorio* 1572: 73).

Rules were closely based on those of conventual orders and discipline was very rigid. Lacemaking in such institutions was certainly no mere hobby or 'occupational therapy'. Each girl's earnings was divided into three parts; two parts went to the Institute for her support while one third was saved for her dowry, whether she wished to get married or take the veil. The girls were each given their daily task by the Mother Superior; if they missed work for no good reason, they were quite mercilessly made to pay out of their subsequent earnings, while regulations concerning deportment and mode of dress were entirely dictated by a dominant concern with modesty and avoidance of shame.[15] For their holidays – usually one day a year – they would be taken on a trip, 'all together on a boat, and a closed one, if possible. They were to behave at all times with modesty and reserve and they would be assisted on their journey by three of the school's deputies on their gondolas (*Derelitti. Capitoli et Ordini.* 1677–1678: 15).

Themes that emerge quite strongly are those of the determination of the state, the merchants' guilds and the convents to retain control over the workers,[16] who often tried to sell their own products directly, thus circumventing the Institutes' strict monopoly over the lace. Since it was in the nature of the craft itself to lead to the creation of networks of women, which were regarded as potentially corrupting if they extended outside the institutions or beyond close family circles, orders and deliveries had to be arranged through the Mother Superior, so as to avoid 'unnecessary friendships'. However, although it was severely forbidden for the women tied to pious institutions to sell their lace, rules were frequently broken by lacemakers eager to offer their handicraft to shops or to Venetian families. An eighteenth-century official thus angrily reports

> the ill-born novelty and the pernicious abuse introduced by several women who go about the town selling haberdasheries...in large quantities and involving considerable capitals. These women sell lace of all kinds white and black veils...needle point...yarns from Flanders and all other relevant goods. All these certainly damage the haberdashers whose capitals are all taken up in the running of shops, while they, with the freedom of their sex, introduce themselves into the homes of nobles, and even into monasteries, where, on the occasion of novitiates, they do sell large quantities of goods (A.S.V. *Arti* B.312 and Gambier 1981: 26).

Women from the peripheral islands, a well known and popular stereotype, as we can see from numerous engravings, figured quite prominently among such illegal street traders. However, as is clear from the passage above, a concern with competition is mixed with suggestions of pollution, and reference to 'the freedom of their sex' implies that, while in being women they were not feared and and could freely sell their wares from door to door, as they did so they were putting themselves at risk. Suggestions that they may have been careless of their sexuality therefore cast a shade on their reputation. Indeed, as I found during fieldwork, street vendors, like gypsies, are always touched with a suspicion of shamelessness and even with allegations of witchcraft (cf. Okely 1983).

The fact that much of Burano's lace was made in the domestic sphere (the cause of a lack of documents, which has dogged researchers' attempts to reconstruct its history) certainly contributed to the survival of ancient techniques. As the women say, their skill was passed down from generation to generation (*de riedo en riedo*) and it was thanks to Burano's distance from Venice that their lace remained unchanged, even when in other areas bobbin lace had become far more fashionable and popular. While convents may have had a key role at an earlier period, a cottage industry probably became the main basis for the island's lace making, when, in the late eighteenth century, several convents were closed down, or lost their prestige and their dominance.

Relations between domestic and convent production were summed up for me at the time of my fieldwork by one of the fathers of Saint Francis of the Desert, who described the lacemakers as 'those good maiden-aunts who used to live and work in the shade of the Church'. While vivid images of *pizzocchere* and of spinster-house-nuns in the islanders' memory confirm that by the eighteenth-century lacework was mostly conducted as a cottage industry, its production was nonetheless tied with convents and religious institutions. Evidence I found from one of the few extant documents, shows that the craft had far greater continuity than is sometimes believed. It may also be argued that the relation between work conducted in Burano's convents and cottages may have differed from those that obtained in Venice. To posit a clear-cut separation between institutional and home production for Burano may even be misleading, given its physical and social configuration and the relatively greater permeability of public/private, institutional/domestic spheres.

Social Structure and Poverty in Eighteenth-century Burano

Evidence of a dominating concern with sexual honour in the organisation of lace making also emerges from a set of papers (to my knowledge never cited by historians of lace) I serendipitously found in one of Venice's archives. My library research – admittedly sporadic, as time was dedicated primarily to fieldwork – was mostly prompted by questions raised in conversation with informants. In this instance, what mainly led me to look at the archives of the city's pious institutions, IRE[17], was the fact that several people in Burano, in recalling their past poverty, spoke of such institutes in ways which implied that these were an integral part of their experience and conceptualisation of their social world. The *Pietà, Zitelle, i Frati,* or 'the Friars', implicitly San Francesco del Deserto, often entered peoples' ordinary conversation, and their practice and skill in applying for assistance, or extending charity and offering help to others when in a position to do so, all implied that behind such familiarity must have lain a long tradition and a long-standing *mentalité*.

A few documents in a slim file revealed new details on the history of lacemaking, on life in Burano in the mid- and late eighteenth century and, above all, on the relation between that island and Venice. The documents, a number of letters, drafts of a statute for a charitable institution, and initial drafts as well as a definitive version of a will, were also an interesting testimony to the life of a distinguished Buranello, the priest Francesco Rossi. His origins were probably humble, but, given that the Church was one of the few avenues of social mobility and that some of his relations also had ecclesiastical careers, his family may have been among the relatively better off. His career as a priest and a member of the Order of St. Philip Neri, the *Oratorio dei Filippini* [18] must have been successful and varied; for a time he was the head of a seminary in Brescia, where, thanks to a naturally frugal nature, he saved most of his earnings, a handsome sum which he eventually invested in Venice's stock exchange, or *Zecca*, to leave to the poor of his native island.[19] On returning from Brescia, Francesco Rossi resided in the house of the *Fillipini,* in Venice's '*Campo Alla Favà*', where he died in 1763, but he never lost contact with Burano, and in his later years he appears to have been a member of a minority of educated and caring persons who concerned themselves with the welfare of their fellow villagers.[20]

To obtain the approval, or at least the permission, of the state authority to establish his institute, Rossi needed to use all his diplomacy, given that by the 1750s the Venetian government was determined to stop the passage of goods and fixed rents to the Church. Indeed, the Republic's political leaders, who had long aspired to limit the influence of Rome and bring the Church under the Doge's jurisdiction, saw all Church property as Venetian patrimony, which, once given to convents and devotional schools, never reentered the economic cycle and remained frozen in 'foreign', that is, Roman, hands (Marcolini 1989: 122–24; and Goody 1983: 48ff). In the same period, Venice was becoming entrenched within its city borders, and it was increasingly difficult for the periphery's impoverished inhabitants to obtain any help and assistance from the centre.[21] For example, admittance to the Venetian Institute of *Zitelle* was only granted to girls born in Venice, and not outside, '*not even in the Dogeship*' (my italics), a rule which automatically excluded Burano's girls.

To preempt objections that might have derived from complicated relations between Church and State – but also, and not least, between Venice and Burano, Rossi makes clear in his application that the Company he was about to found was not going to contravene any of the government's regulations. It was his hope merely 'to promote the service of the Lord...forming an entirely *lay* Company, under the title of *Patrocinio delle Donzelle Periclitanti*'. The Company would include no more than five members, some of them patricians, and it would be a secular one, instituted only to administer a charitable trust. Above all, it was not the Company's intention to buy real estate with a view to starting any hostels or hospices, or constitute a brotherhood, pious place, conservatory or Congregation...but to operate entirely privately with those capitals that will be freely offered.' (1748: 2)[22]

Among Rossi's papers, and contrary to his statement that no real estate would be involved in his plan, are the deeds of a house in Burano's square, which he had purchased and put in the name of the parish priest – a possible indication that his initial plan may have been to found a home as Burano's answer to Venice's *Zitelle*, and that, due to the government's hostility, such a home was never founded, but the premises were let and the revenue paid into a trust fund.

A comparison between the rules for admission in Rossi's drafted statute and those of *Zitelle* shows significant differences between Venice and Burano, where poverty was all the more severe. Rossi's trust, as he emphasised, would dispense with the requirement of an initial payment, which, for the *Zitelle*, was one hundred ducats and a considerable trousseau – a rule which, in his view, excluded 'those who [were] in the greatest danger, because of their extreme poverty which is a temptation greater than any other.' His trust would support the very poor, those 'who do not have the means with which to *cover* themselves with the clothes and the funds required', so that 'they may be taken in with whatever, little or much, they may possess, and, even if they were completely destitute and deprived of everything' (*Supplica* 1748: 2).[23] Their clothes would be 'all the same, durable, modest, and suitable to the poor condition of the daughters, also to prevent vanity, the usual incentive of sex, which leads them astray from the good path'. (*Patrocinio delle Donzelle Periclitanti. Statuto*: 31. A I V, PER A1).[24]

Another difference between Father Rossi's *Periclitanti* and Venice's *Zitelle* was the latter's emphasis on the fact that the girls should be not only orphaned and poor, but also strikingly beautiful, and consequently 'in the most extreme danger of being lost...' After all, as the *Zitelle's* regulations argued, those less favoured by nature could safely be placed into service, while to take them in the institute would have been tantamount to depriving 'such conspicuously beautiful and endangered Virgins' for whom 'it pleased the Divine Majesty that this House should be founded, to free them from the World, from the Devil, and from eternal damnation.'(*Zitelle, Ordini* 1771: 8–9). By contrast, and with greater adherence to Burano's economic reality, the *Periclitanti*'s requirement for admission was that the girls should simply be 'of some liveliness and attractiveness, and of ages between fourteen and eighteen' (*Periclitanti*: 1749: 28).[25]

From the point of view of Burano's social history, the most interesting document is Father Rossi's draft of his will written in 1757. Compared with his earlier writings, the will is far more direct and personal because, as well as expressing growing concern for the 'little daughters, those orphaned or destitute girls' whom he wanted to shelter, it included small bequests to members of his family. In a first draft he left to his nephews his share of the family house, as that was inalienable property, but, as we learn from a later codicil, he also had two nieces, to whom he left a few items of linen and clothing, as if following an afterthought or a reproachful reminder.

In the main, though, he disposed of everything he had earned (but I have found no precise figures) independently of any family consideration. His idea of donating his patrimony for the protection of young women must have been long-standing, as his diverse plans and legal provisions appear to have been the result of careful elaboration over a long period of time. In 1749 the trust supported six girls, but by 1757, when Rossi redrafted his will, his concern for Burano's poor must have been ever greater, since his initial grant of ten *soldi* per day for each girl was changed to the lower sum of ten *lire* per month, so that a larger number might benefit. What is more, while at first the girls in the care of the Company were to be of ages between fourteen and seventeen and be orphaned of both parents, he later understood that the age of girls in need of protection was as low as seven. Such changes show awareness of an ever greater need, due to the island's gravely deteriorating economy.

Indeed, while Rossi was primarily concerned 'to bring the little daughters those spiritual benefits which are of so much glory to the blessed God', he was also well aware of their more immediate and basic needs, had they been entirely homeless and starved. One of the worst ills was vagrancy – a grave problem for both the government and the Church. He therefore made it a condition that his young beneficiaries should be living securely at home, or in the home of some relative or benefactor. He was painfully conscious of the fact that relatives, and even parents, could not always be trusted, and, in so far as a small grant would benefit also the families, he pressed them very strongly to shelter, look after and feed the girls. Through Father Rossi's various drafts of his will, we thus begin to see features of Burano's social world. Perhaps somewhat simplifying, this appears to have been divided into roughly two unequal parts: those, who, like Rossi's trustees, a few village notables, members of the Church and *maestre,* were comparatively prosperous and secure and those who were destitute and hungry.

Rossi's repeated references to *maestre* – and that does imply lacemaking teachers – leave us in no doubt as to their existence and their prominence in the island's society. For example, as he writes:

> It is [our] desire that the girls be sent to those schools where there is taught, with zeal, the holy fear of God, as well as those crafts which are the most usual in the village, above all, the work they call 'of air'. If the girl, or girls, should be in such dire need as urgently to want some sustenance, 'the *maestra* will provide something from her own kitchen or store. For that, and for some little fire she may have to light in order to warm the girls in winter, she will be compensated by the Trustees.

Such *maestre,* therefore, emerge as a well-defined category of skillful and trustworthy women, already well established and known, whose tasks went well beyond those of teaching the girls *punto in aria.* Their financial responsibilities were considerable, since it was they who would distribute Father Rossi's alms to the most needy pupils and every six months they would deliver the money earned by the girls to the trustees. Above all, it was their moral responsibility to see that all the charity's conditions should be obeyed by the girls and their families. As well as being reimbursed for any expense they might have incurred, each *maestra* would receive, 'one silver ducat as a small token of recognition for her trouble…on the first day of the year.'

Finally, we are given a glimpse of Burano's more fortunate girls, as we read Rossi's rule for assigning to the poor grants of ten ducats each, to be distributed by the *Veneranda Schola del Santissimo Sacramento*. Knowing that the number of applicants was always greater than that of grants available, and partly due to Venetian tradition, in which luck was always a factor in allocations of subsidies, grants and even clerical and political offices, a choice would be made by the drawing of lots. 'The grants shall be drawn out of a box with a golden ball by the nubile girls of Burano, of ages between 18 and 24 and no older, in the presence of the guardian, and with the assistance of the reverend parish priest.'

Putting such separate pieces of evidence together, it is noticeable that, over about fifteen years (1748/49 to 1763) in which they were elaborated, Father Rossi's plans show several changes obviulsy related to changed circumstances. According to his earlier writings, the charity was to be run from Venice, and the girls of ages fourteen to eighteen were to be placed in convents, but according to his later will and his codicils, the trustees would be based in Burano, and the age of the girls would be lowered to seven. However, the most important change was the shift of the charities from the convent to the domestic sphere. Granted that this may have been due mainly to the government's opposition to any increase of Church property, the picture of the type of aid proposed is in many ways quite advanced – one in which the teaching of a skill and moral improvement are brought together under the domestic roofs of Burano's *maestre,* as safe havens and as the best settings for the prevention of infantile vagrancy and prostitution.

Compared with Venice's great Hospitals as 'total institutions', Rossi's project, and especially his attempt to shift the burden of responsibility to the community once the state had shown itself hostile to the foundation of a house of shelter, represent an interesting innovation – and one particularly well suited to Burano's social and residential structure.

From the Nineteenth Century to the Present

At the beginning of the nineteenth century, Venetian lacecraft was described as 'languishing' and in a 'state of decay'. In the same period, due to a Romantic interest in folk culture, the lagoon's peripheral islands, like their rural equivalents, began to be looked upon as repositories of archaic skills that had almost fallen into oblivion in urban and less isolated areas (cf. Filiasi, above). References to Burano's *punto in aria* began to appear in the press, and lacemaking in general was considered significant enough, in economic as well as cultural terms, to be mentioned in government reports and statistics. According to the Venice Chamber of Commerce, two-thirds of Burano's women made lace, but, as another writer points out, with greater awareness of the difficulty of forming an accurate picture of the craft, 'the number of workers cannot be stated with any degree of certainty, since, due to the absence of any "positive manufacturer", almost all the women work by themselves at home'.[26] As appears from another report written in 1811, 'No formal factory of lace is found in the Adriatic. Several thousand women, spread through Venice, Burano and Chioggia practice that craft which is said to be in decline'. Yet, the writer continues, 'Burano lace has a great name since it is worked by needle, a very distinguished type of work which deserved the pleasure of her majesty the wife of the viceroy [Eugene Beauharnais] who deigned to purchase some.' (ASV Archivio Camera di Commercio, B1, III. Quoted in Gambier 1981: 31).

During the Italian Wars of Independence (1848–1866) the stagnation of trade and the economic decay of Venice led to escalating poverty, and by 1866, when Venetia was joined to Italy, the city, and to an even greater extent its island periphery, had a vast population of registered poor. When, in 1872, the lagoon was entirely frozen and fishing came to a complete halt, Burano sank to a point of extreme poverty and many suffered from severe malnutrition.

A member of the recently formed Italian Parliament, Paulo Fambri, then started a manufacture of fishing nets amongst the men, but, after that completely failed, he took up an earlier suggestion by Venice's *Podestà*, Count Marcello, that the best way to help Burano was by reorganising its lace workers. An appeal for funds was launched nationally through the press and eventually the Lace School was founded (Pasqualigo 1887:45–9). As Fambri later recalled, 'its organisation was placed in the hands of a committee of ladies – good, intelligent, rich, high placed, and, if possible, beautiful ladies' (1893), mostly members of the newly formed national élite. Among them were Princess Maria Chigi Giovannelli, and a Venetian Countess, Andriana Zon Marcello (1839–1893), both ladies in waiting to Italy's Queen, Margherita of Savoy, who fully supported their enterprise. Some of the society women involved soon lost interest, but Countess Marcello became increasingly dedicated to the management of Burano's Lace School, which, in her later life, absorbed all her activity.

The whole venture, and in particular the rhetoric which characterised the well-intentioned writings of the School's supporters make a revealing commentary on the snobberies, as well as the pieties, of nineteenth-century philanthropists. Indeed, Countess Marcello's life history itself bears witness to the conservative attitudes that informed her commitment to the Burano Lace School. Like her husband, who had suffered a short period of exile in Corfu because of his participation in the 1848 uprising against Austria, she was dedicated to the cause of national unity, but, while some members of Italy's aristocracy, though troubled over the Roman Question, supported the idea of a secular state, her attitude to the Church was one of unconditional loyalty. When, in 1854, Pope Pius IX, striving to strengthen the hold of the Church, established that belief in the Immaculate Conception of the Virgin was to be made an article of faith, or when, ten years later, he at once condemned liberalism, socialism and rationalism in his *Syllabus Errorum*, Countess Marcello always remained one of his most ardent and unquestioning followers.[27]

She therefore reconciled strict adherence to Catholic doctrine with the political career of her husband who had become a deputy in Italy's first Parliament. In Rome, she took an active part in political and court circles; she was a frequent visitor to the Vatican and she may have contributed to keeping open channels of communication, which eventually helped to bring about the Concordat of 1929, long after her death.

In an article written after her death, her friend Paulo Fambri, as well as praising her 'beauty, intelligence, efficiency and modesty', also recalled episodes that reveal an authoritarian and intransigent character. 'When her husband died, in 1869, she was left, not yet thirty, with no less than seven children, a splendid widow, with the most Greek of profiles; in colour the most Titianesque of Venetian beauties…a Venus of Milo in whose breast had entered the soul of Minerva'. So absolute was her adherence to ideals of sexual purity that 'these too, pushed to excess, could result in sheer cruelty'.

For example, in 1881, when two women anarchists, Jesse Helfman and Sonja Perowskaia were condemned to death, she was asked to sign a petition to spare Helfman's life on the grounds that she was pregnant, the Countess curtly refused. Apparently forgetting Catholic doctrine about the sacredness of the unborn, she explained that had the petition been to save Perowskaya, she probably would have subscribed, because the latter 'shabbily dressed and wearing very short hair, as if to renounce any physical attractiveness, had all the appearance of a holy sectarian'. Indeed, at the trial, doctors and witnesses had proved her to be quite unstained, *immacolata*, but, as for Helfman, less young but unmarried and pregnant, she would not have hesitated to leave her to the law (14–15).

Despite such a testimony to Countess Marcello's ruthlessness, Fambri concludes that she was only hard through necessity, but she was in truth rather tender at heart. She used to refer to the lacemakers as her 'three hundred daughters', and she was convinced that piece work would give them far greater freedom and dignity than full-time employment in factory or workshop. However, if the lace was not perfectly executed, she would cruelly cut it with her scissors, sometimes destroying as much as a week's work. That too, according to Fambri, was a hard necessity due to problems that had been left unresolved for whole centuries: those of Burano's poverty and of the decay of the lacemaking skill. It was a great satisfaction to the Countess to see that since the Lace School was founded, the number of illegitimate births had gone down considerably, and that she had contributed not only to the raising of the material standards of the islanders' life, but above all, their morals.

At first the School, which was opened officially in 1872 with only six girls, was a financial disaster, but by 1875 the number of workers had risen to a hundred, and by 1880 Burano lace was beginning to register some cheering successes at international craft exhibitions in Paris and Chicago. In the School's archives, there still remain plentiful records of orders from the Queen of Italy, Margherita of Savoy, Countess Bismark, a Greek millionaire, Papadopoli, a Russian princess who ordered no less than forty metres of lace, the American department store Nieman Marcus and a few English ladies, in particular Lady Layard,[28] who was also instrumental in providing the lacemakers with good quality English yarn.

In 1894 the school was legally constituted as a 'cooperative' – a name that has led a number of observers and even trade union historians to mistakenly assume that Burano's lacemakers had formed, rather ahead of their times for Italy, a true workers' cooperative (Spinelli 1948: 9). Unfortunately nothing was further from the truth, since the cooperative was, in fact, one of share-holders, and its sole Buranello member was the School's male designer and administrator. The only step towards some formal cooperation among the workers was taken in 1895, when a Mutual Help Society was instituted at the suggestion of two of the School councilors.

Far from introducing any innovative ideas in the organisation of work, the School, it was hoped, would revive the moral and organising principles of Renaissance institutions. The girls would have to work six hours a day in winter and seven hours in summer (9–12a.m. and 3–7p.m.) when daylight is longer. They would be admitted to the School at the age of eleven or twelve, provided they were in good health and of good character, and they knew how to read and write. Above all, it was an essential requirement that they should already have learnt the rudiments of lacemaking at home, as it was assumed that they should start learning under their mothers' guidance at about the age of six.

The School, then, offered them an opportunity to perfect their skill, but it was also a way of keeping them steadily at work for very long hours. As well as strict moral supervision, they would receive religious instruction. At the end of their curriculum they might be granted the title of *maestra*, teacher, but between the ages of twelve and eighteen they would work with no payment whatsoever. When they reached eighteen, they would begin to receive some meagre compensation – as count Marcello commented in 1908 'proportionate to their needs' – and, from then on, the discipline would become less severe. Here one may well ask by what rights such an institution should call itself a school, except that, as Karl Marx pointed out, use of that term was a common device for evading regulations on child labour (1952: 229–230).

By that strange mixture of enlightened and cruel policies which so often characterises even the best intentioned of philanthropists, the School's organisers encouraged women to continue working from home after marriage. Whether it was conducted at the School or at home, work was minutely parceled up and each worker was encouraged to perfect her skill in only one or two stitches, so that speed should be gained rather than versatility.[29] As one enthusiast remarked, the island would become 'a rather singular type of factory' – one in which women could be employed almost on a full-time basis without having to leave their homes and children at all. The benefit this would confer on the population in terms of control over women was always spelled out in great detail, so that, paradoxically, at the same time as they were offered opportunities to gain a measure of economic independence, the belief that they had to be kept in check was continually reinforced, since the School was not only founded to alleviate poverty, but also to counteract the predisposition to sin, shame and dishonour that were held to be its almost inevitable corollaries.

A rhetoric of poverty was exploited on both sides: the lacemakers, driven by need, would sometimes solicit passersby to purchase their handicrafts. Ironically, their product was a very exclusive and expensive item of luxury, but, by that very token, it could easily be dispensed with or substituted by other textiles – and, especially in the late nineteenth century, by machine-made Nottingham lace. The fact that it was purely ornamental, and that, given its superfluity, demand was extremely capricious, was used as an argument by sellers and entrepreneurs to keep wages very low. What is more, given that worthwhile buyers of handmade lace were traditionally the very rich, great emphasis was placed by the School's governors on the 'honourable' and privileged nature of the craft, as well as on the more practical point that an ability to sell the product was almost more crucial than its manufacture, because, as countess Marcello never failed to emphasise, it depended on social networks that Burano's lacemakers or Venetian bourgeois traders could never hope to penetrate.

However, once lace regained popularity in the early years of the nineteenth century, Venetian entrepreneurs started competing to obtain Burano's product, and working teams multiplied. Commercial workshops imitated the traditional school organisation, so that the island's lace continued to be almost entirely controlled by outsiders, helped by efficient local intermediaries, especially the lacemaking instructors and designers, who always cultivated their links with prestigious customers and maintained lasting commercial contacts in Venice.

Throughout the whole of the twentieth century Burano's lacemaking has undergone a dramatic alternation of successes and crises. After a prosperous period which lasted till about 1910, profits began to fall, and when foreign markets were closed during the First World War, the School only survived thanks to generous loans from the Marcello family. Business was taken up again after the end of the war, and by 1926 the number of workers had risen to eight hundred (one quarter of the total female population), but by the end of the 1930s sales began to fall again. Under fascism (1922–45) the School was taken under the protection of the *Ente Nazionale della Moda* and it obtained some Government grants. Unfortunately, lacemaking, like other crafts, was negatively affected by adverse tax laws that made no distinction between industry and crafts, while a 'Disciplinary Service for Raw Materials' instituted by the *Confederation of Industrialists,* made it very difficult for lacemakers to obtain good quality thread.

Thanks to the continued interest of the Marcello family in the welfare of Burano, when relations between Church and State, which had been broken since 1870 because of the abolition of the Popes' temporal power, were restored with the 'Concordat' in 1929, the lace School was commissioned to make a precious surplice to be given by the fascist Government to Pope Pius XI for the Jubilee of 1930.[30] Soon after that the School was charged with making cot furnishings for the Princess Royal Maria Pia of Savoy, while a major piece of work, which some of my informants still remember, was an exquisite bridal veil that was to be the gift of the Senate to Mussolini's daughter Edda on her wedding to Count Ciano.

As an old woman remembered, offering an example of a desire for subversion, which in those years could find no outlet other than blasphemous and deflating talk and song:

> Twenty whole metres of wonderful lace. We worked at it for months, then, when it was finished and pressed, we laid it on the table for the authorities to see, and it looked as light as a cloud...they gave us a vermouth and biscuits...They even allowed us an afternoon off, and we all went for a walk in Mazzorbo, laughing and singing 'ahi, ahi, you've torn my veil, you've torn my ve-e-e-il / oh, cruel traitor, you've to-orn my veil!'...So much work, and just to cover her backside![31]

However, such single commissions were certainly not sufficient to maintain the School, and, while the lacemakers' pay actually went down, emphasis on discipline, obedience and sexual repression, in keeping with fascist attitudes, led to an increasingly punitive atmosphere and to strict enforcement of working rules. Numerous women abandoned the Lace School so that, by 1939, the official number of workers had gone down from eight hundred to one hundred. By the end of the Second World War lacemaking was so lowly paid and closely associated with oppressive poverty that women started to take on other occupations. When, in 1962, the School found itself in financial difficulties, the then President Count Alessandro Marcello handed over its administration to the nuns, who had always been the pupils' supervisors and religious teachers. Although the demise of the craft was generally lamented, the School gradually went out of existence, so that, according to documents of 1973, it appears that by that date the Lace School had already been closed for some time.

The Beginnings of Lace

Most of the lacemakers I met in Burano have little interest in problems raised by historians concerning the early origins and the geographical diffusion of lace (see above). However, occasional remarks and opinions I heard during my long period of fieldwork all contributed to a view of the beginnings of lace which partly differed from its official history. While not attempting to provide an answer (which would, at best, have been speculative and uncertain), and leaving aside all learned irrelevancies, the women focused on aspects pertinent to their local history, and to the history of relations between Venice and Burano, the cottage and the convent, the dominant and the subordinated, or indeed, the poor and the rich. The question was: is lace a 'native' art, or was it imported into Burano by nuns?

That the problem of origins should be raised by the lacemakers at all shows that, in the context of recent social change, as well as asserting their economic rights over their product, they would also like symbolically to vindicate their invention of the craft prior to what they regard as its appropriation by economically strong outsiders, or by monastic and philanthropic institutions. Most scholars, as we have seen, consider lace as a derivative of embroidery, and generally imply that the island women were taught it by Venetian ladies in their convents and palace schools. However, as one of my most articulate informants explained, one of the reasons why she regarded learned discussion of origins as pedantic and irrelevant was just *because* she had no doubt at all that lace must have been made in Burano since time immemorial, and she did not believe that it derived from embroidery. Therefore, partly to undermine the new School venture and point out its irrelevance, women very often implied that *punto in aria* is so deeply rooted in their birthplace and their culture that it can just as easily be learnt from their mothers, grandmothers, or older neighbours. 'Making lace, like fishing', they explained 'can only be learnt from earliest childhood. In the old days, men used to make fishing nets, while we made lace to earn a few extra pennies.' Why else would one of the most delicate stitches be called simply *rete,* net, or *Burano*? That lace must have had its origins in the island, then, was not really doubted and was not considered worthy of too much reflection. Legendary accounts of its origins, told to me by schoolchildren who had learnt them at home from their grandparents, are simply part of Burano's lore; they are sometimes briefly reported in tourist literature, and, while they have to some extent been officialised, they at the same time have been 'minimalised'. Differences in various versions I collected were usually related to the narrator's gender. A twelve-year old girl provided the following version:

A young woman was standing by the sea.[32] She was waiting for her lover to return and while she waited she saw that the seaweed had formed a beautiful woven pattern (*un lavoro bellissimo*) on the edge of the water. She liked the work very much, and when she returned home, she tried to copy it with some thread and a needle: she saw that it was coming out very nicely. She then taught other girls, and they all added something to it and they worked out new stitches.

A boy told me the following, somewhat different, account:

One day some fishermen went off on a fishing expedition. They saw a rock and on the rock was a siren. All of them tied one another up on the mast of the boat except one who wanted to be faithful to the woman he was going to marry. The siren saw that he had been strong and she wanted to give him a gift. She struck the water with her tail, and she thus produced some foam. As the sea foam dried on the rock it formed a piece of lace. The sailor [fisherman?] then took it back and it served as a veil for his bride. It was very beautiful and all marriageable girls copied it.

Both versions are related to the problem of the separation and faithfulness of lovers, and both are indissolubly tied to the marine environment. The first does not explicitly involve supernatural intervention, since the lace, made of seaweed, forms naturally and seems to be generated by the sea itself as a reward for the woman's steadfastness. The second version is curious in terms of its geographical setting; the Venetian coasts, where Buranelli traditionally fish, have no rocks or islets. The legend, a much shortened and altered version of the encounter of Ulysses with the Sirens, near Capri, is in fact tied up with the origin of Naples. It would, of course, be very difficult to ascertain whether it may have diffused through oral transmission, or whether it may have been absorbed from learned sources.

What is of interest, however, is its transformations. Unlike Ulysses, who wisely acknowledges his fallibility and gets himself tied up in order not to give way to temptation, the Buranello sailor defies the seductive power of sirens, which he sees as a challenge, and decides to overcome fear and weakness solely through inner strength. In this way he will assert his virility through restraint rather than through the immediate exercise of sexuality. He shows a kind of bravado that does not rest on a capacity to seduce, but on his fortitude in resisting seduction, and remaining faithful to one woman. This is, in fact, rather untypical of manhood in its most aggressive expressions as it has been described in anthropological literature, except among Sarakatsani, where the unpolluted manhood of young *pallikari* is positively valued (Campbell 1964: 279). Indeed, the story, which has some of the features of tales about the devil's challenges, illustrates an implicit belief in some form of mystical control over men by their good wives or 'promised' women.

As some of the older women remember, lacemaking in the nineteenth century was thought to imply some potential capacity to cast magical spells. For example, when a number of girls worked in a group, they would take it in turns to comb the hair of the others, chanting, 'Come here that I untangle you / I want to tell your future / to guess it all for you'. This not only implies that knots, like those by which lace was constructed, might magically have fallen into one of the girls' hair, but also that disentangling the hair would, by analogy, open the way for divination. The double meaning is derived from the Venetian word *destrigar*, which comes from *striga*, witch. *Destrigar*, which literally means to undo some bewitchment, or *strigheria*, has now lost its etymological connection and is simply used to mean 'to tidy up', while it is the second, and similar sounding, verb, *strolegar*, from *strolego/a*, astrologer, which implies the notion of divination. Such associations of lacemaking with magic are now largely forgotten or suppressed, and, although islanders do tend to give great prominence to psychic aspects of life (as, for example, in explaining illness), they would certainly not subscribe to beliefs which they consider naive, mildly amusing, and simply not credible.[33]

Social Change and Fathers' Authority

The history of lacemaking shows an implicit convergence of religious and patriarchal control over women's sexuality which, in Burano's circumstances, was also control over their labour. The main economic basis for the organisation of lacecraft from its beginnings to the nineteenth and early twentieth centuries was the islanders' poverty; the terms of the women's employment were accordingly based on their lack of any bargaining power and their dedication to their craft was explained as due mainly to their need to supplements men's sometimes meagre earnings. From a purely economic viewpoint, such conditions were partly due to the island's complete lack of agricultural land, because, as was the case in other, less land-poor, lagoon areas, even a small vegetable garden would have been a significant resource or an alterative to fishing.

However, cultural factors were at least as strong as economic circumstances: the prominence of lacework undoubtedly was due to the strong polarisation of genders in Burano's fishing culture, but such gender polarisation was greatly reinforced through both secular and religious influences, which Buranelli now regard as having been exercised mainly from outside their community. Emphasis on chastity and severe harnessing of the body, due to the belief that poverty would make women particularly susceptible to sin, and enforcement of rules of hygiene, continually underlined that Burano's women had fallen behind normal standards, and kept them in a state of humiliating inferiority.

As was implied in the pious attitudes preached by religious leaders, poverty was to be accepted, or even embraced, as a permanent and potentially meritorious state, and expressed at all times in dress and demeanour. Yet there lay the crux of a troubling contradiction; while from a religious viewpoint poverty connoted ideas of goodness and piety, it was nonetheless associated with fear of shame and of family dishonour. Most writings on the craft before the Second World War thus underline the charitable intentions of its organisers, and point out that, far stronger than the profit motive, was their sense of moral mission in keeping young women away from the streets to protect them from vice, destitution, infection, and even lice.

Some of the paradoxes of organised lace work in the nineteenth and early twentieth centuries did not escape the condemnation of social reformers, who write about the 'oppressive way work is supervised…in the name of hygiene … and the pain…of those poor and lively workers, too many in number…closed up for too many hours…the fear of "school or convent anemia", the oppression of competitive instructors, the damage to the eyes' and the obstinate refusal of ambitious lace teachers to the potential advantages of machines' (Cibele Nardo 1938). Naturally, such criticisms were not uncommon, particularly in the left-wing press between the two World Wars, but they were effectively silenced during fascism.

Memories of older lace workers, as we have seen, are often bitter, and since the 1960s and 1970s more varied, secure and profitable forms of employment for women have been viewed with relief as welcome signs of social change and freedom of choice. One of my informants, for example, remembered with positive satisfaction a period of time, near the end of the Second World War, when she had been recruited to work at a chemical factory in a nearby island; it was at that time, she explained, that she had begun to 'open her eyes' over matters of social justice and personal independence. Indeed, women's rediscovery of their capacities to leave behind some of the heavy constraints imposed on them by tradition went had in hand with cultural and legal change throughout the country, and led to much altered expectations and notions of personal worth.

The history of lacemaking shows ways in which the treatment of women in institutional spheres was inextricably tied with their early socialisation in the family. Ways in which change has been, and is, taking place since the 1960s, therefore, clearly parallel change in the exercise of paternal authority, which was gradually softened, renounced, or exercised with remarkable caution. Recalling the atmosphere of protest and indignation that had accompanied the closure of the Lace School in the early 1970s, women and men alike emphasised the political aspects of the long history of women's work in Burano, which, they said, in earlier years most of them had been too simple and too poor to fully grasp. Emphasis on the supposed inclination of the women to bring shame on their families was, as a man said to me 'just a trick' by which the dominant classes had kept them all in subjection. The mere fact that outsiders should come and tell them how to organise their lives and their work, *that,* they said, had been everyone's humiliation and shame.

The history and changing notions of *vergogna,* and of its role in the organisation of women's work, thus brings to light an important dimension of political relations between Venice and Burano. In the next chapter, returning to a historical time closer to the present, I shall further explore political and administrative relations between centre and periphery, and describe ways in which memories of the past, and a rejection of social and sexual shame by people of both genders, has become part of Burano's political discourses.

Notes

1. The exhibition was an extended version of a show in Milan's Poldi Pezzoli museum.
2. Showing the women at work would have been more persuasive to potential buyers than a mere display of old lace.
3. Pasqualigo, the father of Burano's doctor, was deeply attached to the island, well-acquainted with its archives and close to some of the lace workers.
4. It is not clear at what stage references to biblical influence may have become more frequent than classical examples. A parallel with scriptural themes in painting would point to the Counter-Reformation and the latter part of the sixteenth century, but according to Pasqualigo scriptural influence would have determined the character of lacemaking from its very beginning.
5. Biblical quotations are from: A. Cohen. Soncino Press 1947.

6. The biblical expression 'weaver in colours' is usually translated as 'embroiderer' and there is no specific mention of lace, unless we wish to so interpret the Hebrew word *lul*, rendered as loop or tie (*Exodus* XXVI:4, footnote: 507). From a technical viewpoint, the knotted fringes at the corners of prayer shawls are closer to lace than are other fabrics. The textiles described in *Exodus*, are woven and embroidered. Instructions for their making are delivered to a man, the first master craftsman, Bezalel, while women are mostly referred to as spinners and dyers.

7. The image of the veil has given rise to rich symbolic elaboration. In the Jewish esoteric tradition, a veil separates God from all else – even his angels, while, in general, direct visual experience is considered dangerous: women must be veiled in Synagogue, where men too must cover their heads and wear prayer shawls (G. Scholem 1961: 72). In early Rabbinical literature, shame, *Tseniut*, as concealment or modesty is an important moral injunction.

8. Textile crafts were in any case widespread throughout the Mediterranean, the Middle East and the Balkans (Lindisfarne 1997).

9. As M.R. James writes in his critical edition 'we have the book in the original Greek, and in several oriental versions of which the most ancient is a Syriac version, but no Latin version is extant'(1924: 38). It was introduced into Europe by Guillome de Postel in the sixteenth century, and was printed in Venice under the name of *Protovangelium*. Postel was in Venice in 1537 and again in January 1547. It is an intriguing coincidence that from that date to the summer of 1549 he was a chaplain in the Hospital of Saints John and Paul (Bouwsma 1957: 94). Between Postel's two visits, Pietro Aretino published a charming retelling of the Life of Mary which he hoped would help him obtain the Cardinal's mitre (1545), but the book only brought him obloquy, and did not much advance his ecclesiastical career.

10. Another narrative, probably dating back to the eighth or ninth century, is the *Gospel of Pseudo-Matthew* (or *Liber de Infantia*) which was a very influential source of inspiration to artists and poets in the twelfth to fifteenth centuries. Yet another version, known as *The Gospel of the Birth of Mary*, and attributed to St Jerome, passed almost unchanged into the *Golden Legend* of James da Voragine. Other versions are the Coptic Lives of the Virgin (M.R. James 1924: 19 and 38–90).

11. According to Italian popular legend, to convince the unbelieving Thomas that she had actually ascended to Heaven, Mary lowered her girdle from the sky.

12. As G. Lienhardt writes in his book on the Dinka, 'The practice called *thuic* involves knotting a tuft of grass to indicate that the one who makes the knot hopes and intends to contrive some kind of constriction or delay. In one of the texts quoted, for example, an enemy is to be "knotted in grass", meaning that it is desired that his freedom of action, mental and physical, shall be restricted' (1961: 282).

13. Respectively Giovanna Dandolo Malipiero, and Morosina Morosini Grimani.

14. Zitelle, Penitenti, Derelitti a San Giovanni e Paolo, Incurabili and San Lazzaro ai Mendicanti. Not all pupils came from a poor background.

   Linguistic evidence of an association of lace with conventual institutions also derives from the common root of *pizzo*, one of the words for lace and *Pizzocchere*, the name of an order of Franciscan tertiaries, that is, in a sixteenth-century definition, 'women, mostly widows, who having withdrawn from the world, either through devotion or necessity, reduce themselves to live in certain places which are specially set aside for them. Thus retired, they live on alms and on some small exercise of their own'(Cesare Vecellio 1598; cf. Beguines and *Roman de la Rose*). Small communities of such women, usually housed in Church property, were widespread in Venice, where a street in the centre is still named after them; Burano too has a *Salizada del Pizzo*, and, at the time of my fieldwork '*Pizzocchera*' was the ironical nickname of a willful and eccentric old woman whose behaviour was by all accounts a perfect reversal of all religious discipline and ideology.

15. The girls' clothing was to be 'altogether modest…with a simple, young-looking dress and a bodice in the same colour with no frills or pointed ends'. They were 'forbidden to wear lace, silk

shoes and bows, as well as coloured garments, earrings, lockets and gold rings, or any other kind of ornaments'. Their veil would be made of cotton, 'in a severe close weave – not transparent...they will have to dress with that modesty that is fit for the poor girls of this pious hospital' (Derelitti 1677–1678: 15).

16. Lacemakers who attempted to emigrate to France were threatened with the death penalty.

17. Since 1975 those documents have been gathered in the archives of I.R.E., *Istituti di Ricovero ed Educazione.*

18. Saint Philip Neri (1515–1595) founded the Congregation of the Oratory. This was initially a society; members did not take vows and services consisted of a musical composition on a biblical or other religious theme, sung by solo voices and a choir, hence the name *Oratorio.* Neri placed greater emphasis on the values of gaiety, gentleness, integrity and love of God and man than on austerity and physical suffering.

19. As appears from a number of letters in the archive of the *Pio Luogo delle Penitenti di San Job,* an institute of shelter for repentant sinners (1700–1797), Francesco Rossi was the younger brother of its principal, Elisabetta. Documents in the archives of the two institutions testify to a strong communality of ideas among their founders, secular heads and spiritual leaders, who appear to have formed a close-knit network, linked either by kinship, or friendship and spiritual affinity.

20. Most of the people Rossi named as trustees of his charity in Burano were either priests, members of religious orders, or of the School of the Blessed Sacrament, one of a number of devotional societies disseminated throughout northern Italy in the late fifteenth century, mainly to repress blasphemous attacks against the Eucharist. As Pullan points out, the 'Blessed Sacrament' , which owed its inspiration to the Observantine Friar Bernardino da Feltre,was a very humble lay congregation, which, compared with Venice's *Scuole Grandi,* was 'liable to be dismissed as a thing only fit for smiths and cobblers' (1971: 121).

21. At that time, companies of laymen were formed to minimise the extent to which new institutions were backed by the Church. For example, when the institute for repentant sinners, of which Elisabetta Rossi was the principal, was founded in 1705, a petition to open it was signed by 'four citizens'. Its main patron, however, was Venice's Patriarch, Badoer. To show that the institution was a secular one it was decided that the women should not wear uniforms, as 'that might cause the Government some jealousy, as if they were about to found a convent'. The *Penitenti* was able to survive by vindicating its secular nature and declaring its absolute poverty, as Marcolini (1989: 127) shows, not entirely truthfully. It actually continued to take in young women till 1956, and people who remember it still describe the women as *redimende.*

22. During the Crusades, Venice's shelters were mainly hostelries for pilgrims on their way to the Holy Land. In later periods, especially the fourteenth to the seventeenth centuries, there developed a strong bias in favour of the sick and the poor, but, more markedly still, the protection of women. Out of 68 new foundations, 34 were for women only; one of these for noble widows, one for gentlewomen, 6 for *pizzocchere,* or Franciscan tertiaries, and one for ex-prostitutes (Semi 1983: 37).

23. As father Rossi writes 'Six girls [are] already gathered, five of whom are placed in education in the convent of San Vito in Burano... another, due to some weakness of spirit and some other physical defect, is in the care of the Maestra Antonia Fiorelli (1749: 22).

24. The statute contains instructions on book-keeping and rules for the running of the charity. The Cashier would visit the girls once a year, to enquire into their conduct and to discover their intentions for their future. He would have a salaried aid, paid 30 Ducats a year, whose main duty would be to find new girls in need of shelter.

25. Whether the girls should decide to take the veil or get married, 'everything [would] depend on the decisions of the Company'. If a girl should wish to be a nun, she would be found a place where the expense was the least, even if that was outside Venice. However, the Company would meet all costs, and even assign her an allowance of 12 ducats per year. Those who wished to marry would receive a dowry of 300 ducats. The Cashier would take care to find safe matches,

whom the Company would examine 'with the greatest of cautions', and wedding contracts would be signed by all the trustees. As a reward for their trouble, 'on the death of any of the members of the Company, all others…shall pay for the celebration of three Masses, since it seems reasonable that, if the Company takes on as its purpose the spiritual well-being of others, it should…also provide for its own (*Periclitanti* 1749: 33). The Chapter ends with an invocation 'that the Company may be humbly commended to the protection of the Holiest Immaculate Virgin and St. Philip Neri'.

26. As R. Williams (1976) has shown for the English word 'industry', before the Industrial Revolution also the Italian *industria* and *manifattura* designate crafts, or for *industria*, a disposition to work hard, dedication and versatility.

27. An interesting detail is the presence in Burano of the nuns of Maria Bambina (see Chapter 3).

28. Lady Enid Layard was the wife of Sir Henry Austin Layard, who retired in Venice in 1883, after his successful career as an archaeologist, and his subsequent failure to enter English politics (Pemble 1996: 34–5).

29. This rule, which in the past may have been adhered to in the interest of entrepreneurs, is still accepted by the women. They believe that several hands should contribute to the making of each piece of lace, a custom which, now outside the school structure, has led to the creation of small chains of cooperation, enduring friendships and a sense of mutual dependence.

30. The Concordat, which was to confer a semblance of respectability on the fascist leadership, was a diplomatic triumph for Mussolini.

31. *Ahi, ahi, m'hai rotto il velo / m'hai rotto il velo // Cuore crudele, tu m'hai tradi!* The song, in which the veil symbolises the hymen, is usually sung by a man mimicking a woman's voice. As the words say, she complains about her lost virginity. Sung by the old woman, the protest was clearly ironical and amused rather than plaintive. Indeed, she was poking fun at the idea of an imposed virginity and at the pretensions of the powerful, in that case the mockery was against the daughter of the dictator, whom she did not hesitate to qualify as a true *putana* (prostitute).

32. Confusing two different narratives, the girl actually gave the woman a name, Francesca Memo, nicknamed Cencia Scarpariola, who lived about one hundred years ago, and who was the supposed heroine of the nineteenth-century rediscovery of lacecraft. She was called by the School's organisers to teach *punto in aria* as most other women had entirely forgotten it.

33. In the past, fear of magical binding was widespread in Italy, where *legatura* was thought to be operated by a rejected or unsuccessful woman, jealous of a desired and unavailable lover. The spell would render him impotent, affect his health and numb or paralyse his limbs (De Martino 1972).

Figure 7.1 The island of lace. Lace and lacemaking are always emphasised in Burano's picture postcards.

Figure 7.2a Angela Adorni, Head of the Casa delle Zitelle (Media Library of the Veneto Region)

Figure 7.2b Elisabetta Rossi, Prioress of the Hospital of Penitenti di San Giobbe (IRE collection)

Figure 7.3 The body techniques of lacemakers and makers of fishing nets.

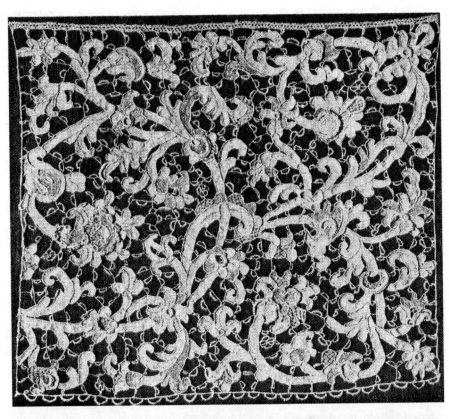

Figure 7.4 Old lace. Two late nineteenth century borders from London's Victoria and Albert Museum (Inv. 347 and 347a, reproduced in Gambier 1981: 57)

Figure 7.5 Contemporary lace by Lucia Costantini. Lucia's use of colour and her introduction of her own patterns and designs are regarded as greatly innovative and original.

8
DEVOLUTION FROM THE GRASS-ROOTS: LOCAL INTEREST AGAINST IDEOLOGY

৵৽৵

The main business of politics is to change things.
(Michael Foot, *The Observer* 4/3/2001)

The Chironomidi

August 1988 was a very uncomfortable and worrying time for the inhabitants of Burano. The presence of huge numbers of tiny, insidious, mosquito-like insects, or *chironomidi,* and an excessive amount of seaweed floating in large clusters in the vicinity of the island's embankments or drifting away slowly on the water mirrors like moving islets, were very clear and alarming signs that the lagoon's pollution had reached dangerous levels. The water's temperatures had risen to 27–28 degrees Celsius at a depth of five metres, while streaks of white foam revealed an unhealthy lack of oxygen, and large numbers of fish were dying.

The high level of pollution in the stifling August heat had affected also the historical centre, as well as Murano, Lido and Giudecca, but, perhaps due to the direction of tides and winds, as people there assured me, Burano's plight was by far the worst: first the pungent smell of seaweed rotting under the hot August sun had plagued Buranelli to a greater extent than other Venetians, then swarms of *chironomidi,* themselves the biological product of algae, had come upon the island in such numbers and with such persistence as truly to recall one of the plagues of Egypt.

In the description of several informants, *chironomidi* are 'small black insects, which in the evenings swarm into the island in streams of thousands and

191

millions'. They are attracted by the lights and by any bright coloured surface, so they enter the houses with incredible speed and with unfailing sense of direction. People soon started to bolt doors and windows, but the insects were so small and so plentiful that even the tiniest chink was enough for them to penetrate and cover entire walls and window panes; they crept under people's clothes, they got into their hair, ears and mouths and they completely blackened their television screens. Indeed, as the children complained, 'you can't even go near a tree during the daytime, because if you touch a branch, or a leaf, dead chironomidi come down on you like a downpour of black rain'.

Under those circumstances, Buranelli stressed, their concern over ecological damage was not merely an academic or political matter, as changes in the water and air had fully shown their frightening nature and the extent of their disrupting effects on peoples' everyday lives. Disturbed in their sleep and shocked at the intensity and persistence of the insects' invasion, Buranelli at first entered a dispute with city experts on the best ways to keep the insects at bay, then disagreements soon turned to recrimination over past failures and, in the end, to polemics and open accusations against the municipality over the general neglect of the needs of Burano. Long-standing antagonisms were temporarily echoed and actively revived.[1]

The situation on Sunday August 14 was described by one of the islanders in highly dramatic tones in Venice's daily newspaper:

after 9p.m., when the sky grows dark, and when the canals full of seaweed change from their rotting brownie-green to the black colour of night, the island enters an absurd dimension. Despite the torrid heat, the windows are closed. All the island's street lights are covered and Burano lives its evenings and its nights under a curfew, as in time of war. Most people remain indoors, while a few men in overalls are running around carrying pumps, their heads covered with helmets. That is not the set of a horror show, nor the end of 'The Day After' [the film which showed the after-effects of an atomic explosion]. It is what is actually happening in Burano…The cause are the *chironomidi* …The men in overalls are trying to fumigate the island, but, as its inhabitants point out, in spraying different neighbourhoods on different nights, they are making a mistake, because instead of exterminating the animals, they just cause them to move from one area to another. Who knows why, when there is an election meeting, politicians arrive by the dozen, but when it is a case of sending someone to help, only three or four men are available. (*Il Gazzettino*)

Echoing the islanders' complaints, a Venetian reporter blamed the invasion of insects on a series of failures on the part of the city, and she thus gave way to a polemical exchange which rose to a high pitch over the following days. A second article by a local man, on August 15, begins:

God helps those who help themselves.' *Chi fa da sè fa per tre.*' Buranelli have little hope that help may come to them from the historical centre. They have decided to organise themselves…The most ingenious was the parish priest of San Martino, who on Friday evening lit a large beacon on top of the bell tower…and it disappeared behind a black wall of flying midges. The chironomidi plunged in their thousands and covered all its four sides (72 meters high), as well as the wide façade of the Church. The Piazza and its immediate vicinity were thus given a breather, while the rest of the island suffered another evening in darkness and isolation. (*Il Gazzettino*)

During the following days, while the invasion continued, ecologists maintained that the priest's beacon would just attract more and more insects. To divert their course away from the houses, and following a method which had proved successful in one of Venice's neighbourhoods, a boat was moored in the island's vicinity and covered with white sheets rendered luminous by a number of light bulbs behind them. Meanwhile the *chironomidi* had invaded the historical centre and there was little hope that they would let up as long as temperatures remained high, but their attack on Burano was still the fiercest and antagonism against Venice and its city council continued to grow.

Ironically, the Venetian councilor for ecology was on holiday in Sardinia, a fact that inevitably attracted hostile and sarcastic comments and prompted the Mayor to offer to visit Burano in her place. Having arrived back at the last minute, the councilor for ecology undertook to go to the island herself, but, whether through error or deliberate choice, the president of Burano's council waited for her in vain in the Piazza for almost two hours and their meeting never took place. Resentment continued to grow, and, as all attempts at mitigating the ill-effects of the insects' invasion met with little success, differences of opinion on how best to cope led to a great deal of controversy.

The Buranelli proposed that all street lights should be turned off, but Venice's health authorities feared that total darkness might itself have been a hazard; for example, some child, older person or tourist might have fallen in the water, and who knows what ill-intentioned people might get up to under cover of darkness. To reach a compromise, it was agreed that only half of the street lights, and especially those nearest peoples' front doors, should be turned off. Meanwhile, the municipality had decided to try using yellow light bulbs, which were thought to be more effective than ordinary white ones in keeping the insects away. The Buranelli's question then was, 'why had these not been tried out before, given that they had already been around for a few years?' In any case, they argued, they were not satisfied; the yellow light bulbs were just a palliative, and what was needed was a radical solution, of which no prospect was seen. In answer to their protest, the number of men in protective overalls sent by the commune to spray the walls with some insect repellent had been increased from four to eight, but that too had proved ineffective. Then people, as ever feeling that they were missing out on the advantages of modernity, demanded that the entire island should be sprayed from a helicopter, all at once and not one neighbourhood at a time, but that request too could not be granted because such full-scale spraying might have been a hazard to health.

✳   At the same time, all was not well at the Venice city council. The Mayor was very critical of current environmental policies that he knew were inadequate, 'they are only palliatives, and they only deal with effects, while it is the causes that should be removed'. In general, he took a very pessimistic line, 'All we can do is take some temporary measures: set up lighted white sheets, change the colour of the lights...But the *chironomidi* are here, and now we have to put up with them'.

In sum, as he stated in his address to a meeting of mayors from Italian and Yugoslav Adriatic coastal cities, while he hoped that their joint action would help exercise stronger influence than isolated requests, he warned them not to work up too much hope that major problems, such as the clearing of pollution in rivers, would find any prompt and definitive solution. Damage wrought over years of polluting industrial activities could not be repaired within a few days. Something had to be done to alleviate the consequences, but if attitudes to the environment were not radically changed, it was not within the power of governments to clear away pollution. It really was a matter of education and respect; everyone was responsible: 'those who threw trash into the water, those who used too much fertiliser, the advertisers who pushed polluting products...For too long ecological needs have been sacrificed to development ...but there is no *real* development without environmental protection'.

He recalled that the Special Law for Venice was issued after the workers, in a great political demonstration, affirmed that the defense of the environment was no less their right than was full employment. Indeed, considering the damage to the quality of life (and even to the tourist industry), it was not true that polluting was less costly than its clearing. And, if that was so, people simply had to adapt themselves to living more modestly; it was 'a problem of civilisation.' Contrary to a journalist's allegation that he had not been sufficiently concerned with environmental problems, he recalled that, on the contrary, when he was councilor for the environment (the first in Italy), he had even been threatened by trade-unionists because of his objections to industrial development. He was very well aware of damage to the ecology as far back as 1971, and in 1973 he had resigned from the council to protest against the general indifference to ecological issues.[2] 'Today I find myself having to manage an emergency due to causes I denounced fifteen years ago, and which I too have to bear both as a citizen and a Mayor.'

At that time, however, the Buranelli, exasperated by the insects' invasion, had little patience for the Mayor's long-term view. Their requests for help from the commune multiplied, and they regarded the fact that their requests found little response as a sign of hard-heartedness and of a lack of concern. Claiming that only protest on the title-page of the newspaper had induced those in positions of responsibility to produce those yellow bulbs which were supposedly insect repelling, but had been left unused in a store-room for years, the Venetian journalist again added fuel to the islanders' resentment. Among endless recriminations, also the then president of the *consiglio di quartiere* made the Buranelli's feelings known through the press:

An A B C on the *chironomide* would certainly be a best seller in Burano... People are so incredibly tolerant that they would be ready to live with the insect if only someone could teach them how to do so. Indescribably patient first with the seaweed emergency, and now with these insects, Buranelli have always been the hardest hit; they have never made a protest that was not within the dictates of good manners and respect of authorities – despite the fact that the latter seem to be hiding away from them and absconding. The Buranelli now say that, seeing that the chironomidi cannot be got out of the way, the city's environmentalists should organise courses to explain their habits. If it were to be the case that chironomidi love blue and hate yellow, we could certainly change the colours of the houses. (*Il Gazzettino* 17 August).

The *chironomidi* crisis, however, also gave him occasion to express the Buranelli's long-standing resentment about their treatment by the city on almost every current problem or issue.

We have always been third class citizens...within a few years the island will be as bad as Venice, if not worse: more than half of its population will be old, and all the young will have emigrated...So many promises, but we are still without a health centre, without adequate public transport. For years we have asked for help, but have obtained nothing. Those white sheets now used against the insects have been abandoned for years, they have not been soaked in disinfectant and have not been flood-lit. Here the seaweed is collected by the fishermen on their old vessels, while around the Lido (who knows why?) the commune have made available all the most modern boats. Four months ago I illustrated to the ecology councilor the situation in which we feared we would have found ourselves this Summer, and the answer was not to worry, everything was under control...When will they consider us as good as any other Venetians?... A hard question, because the problem of second-, third- and even fourth-grade categories still exists also in the city, where it would be less difficult to solve than in Burano.[3] But not even that has been started. Just imagine. (Ibid.)

After the troubled summer of 1988, when the Buranelli fully realised the extent to which damage to the lagoon could affect their lives, as one of my friends in Burano recalled (not without a touch of satisfaction) they almost caused the city's communal junta to collapse. To answer repeated charges that the commune had not done nearly enough to help deliver Burano from the insects' invasion, the Mayor of Venice at last decided to attend a meeting of its district council. On that occasion too, the president of the council enumerated the main problems besetting the island, namely, adverse building regulations; pollution; the need for new medical premises and for a gymnasium and, in general, sports' facilities for the young. Finally, and not least, the municipality had done little or nothing to ease or solve the problems posed by high tides...so that, at their peaks, water entered the houses.

As soon as the president's report was over, a member of the audience asked to speak:

At long last we know that we have a Mayor. We understand from the newspapers that you are a caring person. I think I am a good person too, but when I speak with Venice's councilors, I get the impression that we Buranelli are treated like second-class citizens. It is always us who have to go to Venice. Venetians do not have to come here to pay their taxes, get identity cards, certificates, papers and so forth, you name it. Now, just because we are off centre, we need to keep our contacts with the hinterland. [Numerous Buranelli own cars which they park in Tre Porti, about fifteen minutes away by boat.] Now you have granted building permission for a lot of expensive garages on the sites where we have always parked our cars. But that is the same as encouraging another exodus. We cannot pay 28 million lire to buy ourselves a little garage!

This council wants to join us by a terminal to Ca' Noghera, but, as I have maintained for the last twenty years, and as I still do, Tessera is a better place. The general plan is all wrong'. (Map 1.1).

At that point, as the president called the speaker to order, the latter complained that he was always stopped talking and appealed to the Mayor to listen to him, but, as the president readily answered, 'I have not asked you to stop, but to stick to the subject', laughter brought about the desired effect.

A second councilor complained that inland communes received more generous financial backing than Burano. After him a socialist councilor took the word,

Thank you, lord Mayor, for coming. For thirty years we have fought together to solve the problems of Venice. Of these we have more than our fair share. Today Burano has nothing to envy San Marco. We are Venetians like all other Venetians. You are a person of rare sensitivity, you feel the citizens' pulse. Then listen to our requests of at least twenty-two years standing. Our lagoon is polluted beyond all toleration. The copper utensils in our kitchens, the silver, for those who are lucky enough to have some, even the gold teeth in some peoples' mouths, are getting a funny colour…Something must be done…never mind all your great committees [*comitati, comitatoni*]. We must clear out the pollution. And why must we live on ground floors, when in Venice these are used just as entrance halls?

I do not want to be a *qualunquista*,[4] but, as concerns the country's institutions, we have arrived at a point of no return. Unless something changes within a year, as politicians, we all stand to lose: people will just not come to vote. At that point Burano will be the best thermometer to test the state of the country. The health service has not yet been organised… At least in Africa they have a few missionaries…Please, take careful notes of what I say, and pass it on to your functionaries: in Burano, we are honest people.

Several councilors, then, took turns and stated their views on their different areas of knowledge, housing, health, pollution and so forth. Finally spoke a prominent Christian Democrat councilor:

We must make manifest our state of mind. Our *consiglio di quartiere* has not been able to become an incisive interlocutor of the Venice Commune's. We have not succeeded in developing a viable relationship with any of its administrations. We had asked to hold a meeting with the whole communal junta, because problems should be discussed in a spirit of collegiality. It is very good to have the Mayor here, but it is ultimately the whole of Venice's council that has to solve all problems.

In our local council, we certainly have worked above and beyond politics for the good of Burano, but, when we send in a request to Venice, we get no response, not even a negative response. The junta keeps changing.

Like the previous speakers, he talked of the need to have easy access to the hinterland and to keep the lagoon clear of pollution, especially since one fifth of Burano's population actually earned their living by fishing. As he went on speaking, his voice grew louder and louder in a crescendo, till at the end his message was shouted out, as if from boat to boat across stormy waters.

The lagoon was created by the Serenissima. Venice's managers must begin to make serious choices. And that goes for Burano too, as for all other districts. But sometimes we are not even consulted. What we are asking for now is a dialogue with the whole junta, that is, all of Venice's councilors, who are experts on different problems. Not just with you, Mr Mayor – much as we like having you here.

As the Mayor began to speak, the gathering fell into absolute silence. He first of all thanked very warmly those who had invited him to Burano, and he assured his audience that he was well aware of their problems. However, a meeting with the whole junta, he thought, would not have been very productive. It might be far more useful to hold specific discussions with experts. He was very glad to be there, but did not think it a special thing that he should be thanked for. He had planned in any case to have direct contacts with Pellestrina and Burano – both areas with their own autonomous and different characters. 'All districts think that they are uniquely badly treated, but things are being done'. A general plan was nearly completed, and a gymnasium was about to be built (figure 8.1).

He praised the Buranelli's skill and concision in enumerating their problems, and, above all, he admired their *passion*. As Mayor, he could not make absolute promises, but, as a person, he certainly could foresee that some agreement would be found, although, on many issues, the last word was, of course, that of the city architects. Concerning environmental conditions – and he himself had been a councilor for the environment back in the early 1970s – as he tried rather helplessly to convey, damage had harmed the entire lagoon; it was not only Burano that was affected. Again he assured everyone present that efforts would be made to control the pollution that had brought about such an exuberant breeding of algae and thus prevent *chironomidi* developing to intolerable proportions.

As the meeting was closed, with all the customary courtesies, the Mayor took leave, while the men began to wander away amid a continuous alternation of voices, loud whispers, shouting and laughter: the sounds of age-old diffidence and skepticism.

For Venice's administrators Burano's needs were not really different from those of other estuary islands and marginal areas; and, although in earlier years a great deal of attention had been given to its housing problems, by the late 1880s the island did not enter Venetian debates as a specific and separate issue. The *chironomidi* crisis, especially given that it had affected Venice as well, at first appeared a relatively minor and short-lived affair, but the way in which it was discussed and treated highlights the interdependence of part and whole – indeed of different and overlapping political and administrative entities: the neighbourhood, the city, the region and the state – as well as, albeit indirectly, the influence and relevance of global politics and changes in world order.

Debates in Burano's council echoed a general rethinking of political ideologies that radically changed Italy's party politics after Glasnost. Furthermore, due to processes of devolution and representation mainly supported by the left, throughout the 1970s and 1980s the country had undergone considerable social change. After the heavy protests of the 1970s, and the development of a greater awareness of their civil rights, Italians in general seemed to have turned to debate and negotiation. As Putnam emphasises, a sense of civic responsibility had acquired new vigour and there was a growing awareness of social and administrative issues. In roughly the same period, however, northern Italy had witnessed the rise of several separatist leagues, first among them the Liga Veneta, formally founded in 1979, but prepared through cultural activity throughout the 1970s.[5]

The Liga was in fact a populist, anti-intellectual ethno-regionalist movement, which laid great stress upon a northern Italian work ethic, creating a distinction between a supposedly 'European' Italy and a 'Mediterranean' Italy, with strongly racist and antisouthern pronouncements. Although the Liga had a strong following in the Veneto's inland provinces, especially Vicenza, Treviso and Belluno, and to a lesser extent Padua and Verona, it found little response in Venice. Buranelli also showed little enthusiasm for the Liga's politics. Despite the complexities of their relations with the commune, they strongly identified with the city and generally followed its cultural and political orientations. However, although the Liga had little following in either Venice or Burano, it often led by example to strong pronouncements against the central state and lent new vigour to expressions of exasperation and pressing demands.[6]

Aspects of Burano's relations with Venice also emerged during the campaign that preceeded the 1990 administrative elections to chose a new Mayor and new regional, city and neighbourhood councils, when Venetian representatives came to illustrate their party lines and present their programme. At that time opinion was sorely divided over a proposal mainly supported by Gianni de Michelis, one of the candidates to the city's mayorship, that Venice should host a massive world exposition, the EXPO 2000. As well as putting forward various measures to improve the Buranelli's lives and ensure their continued residence in the island, politicians generally presented their views on the proposed exposition and they discussed projects for the whole of the city, trying to involve Buranelli's opinion in matters of common concern.

The May 1990 Administrative Elections

As we have seen in Chapter One, political debate in Burano always revolves around the islanders' practical needs while ideological differences have to be brought into line with administrative problems. From the point of view of the city, these are regarded as relatively simple and soluble ones, but, as I found during fieldwork, that was certainly not the view of Burano's inhabitants. Some steps had certainly been taken by the municipality, especially in the area of housing, but once the pattern of neglect and social distance from the city was broken, their desire to continue bargaining for better conditions had become all the stronger. Choices of candidates for *consiglio di quartiere,* therefore, were not based only on party or ideological loyalties, but on judgments as to who might have been the better brokers and representatives of Burano towards the Venetian municipality.

The topics raised by the island's political representatives only partly coincided with those debated in Venice. For example, issues of cultural identity were not experienced in the same way, for, while Venetians were beginning to feel that globalisation and predatory tourism were a real threat to their collective identity, Buranelli were assured that they were strongly marked through their isolation and they saw all their problems as more specifically relational and practical ones. Debates relating to large-scale institutions, as, for example, expansion of the university or the finding of a permanent seat for the city courts, clearly appeared to be mainly the concern of the centre, but those to do with environmental protection and with the care and control of the lagoon were viewed by Buranelli as alarmingly relevant to their continued life in the island. The problem of housing was one they shared with many Venetians, but solutions were in fact close at hand, and a number of families had already been satisfactorily resettled in nearby Mazzorbo. Nevertheless, regarding social welfare, health services and transport, the islanders still felt very keenly that their position was that of a neglected and disadvantaged periphery.

The election campaign in Burano, especially over the last few days before the vote, due to take place on 6 May, was conducted with some of the noise, excitement and confusion of a popular *festa*. Speakers' platforms were erected in front of the Christian Democrat and Communist Party's headquarters, both in the main street, Via Baldassare Galuppi, and just a few yards away from each other. Representatives of different parties took turns to speak in rapid succession. They had to shout their messages, sometimes over the heads of passing tourists to a somewhat indifferent audience, mostly made up of men and a few women, scattered in small groups, and often conversing among themselves.

Local members of the various parties acted as hosts to speakers from outside; they usually introduced them, and in the end they added their own statements on the needs of the islands and their plans for government. Some speakers arrived on the public boat, and so great was their eagerness to collect a trail of supporters that they persuaded other passengers to join them and attend their address. That, however, was not the case for the socialist candidate to the mayorship, De Michelis: due to his *presenzialismo*, his plan for that one day was to appear in Pellestrina in the morning, then visit parts of Venice and reach Burano in the early afternoon, after stopping for lunch at the Cavallino.

As he arrived, landing from an expensive speed boat – a large, portly, and yet swift figure, accompanied by several of his party's members – he was met by a few local socialists, one of whom, carrying a huge pack of red carnations, handed them out to anyone within reach, until there formed a small procession of people, headed by a few notables and by De Michelis himself. As they directed themselves towards the Piazza and the main street, one could not fail to be reminded of the arrival in Burano of its main patron saint, as it is naively depicted on a large panel in the island's parish church.

'De Michelis's meeting with electors', advertised on large posters, was to take place in the main restaurant, Romano. While he lingered briefly outside and several people demanded to be photographed with him, he was ready to accede to their request, putting his arm round some unknown woman's shoulders with easy familiarity. After a walk up and down the main street, and a brief visit to a shopkeeper of known socialist views, De Michelis entered Romano's. Followed by a sizable cortege, he crossed the large dining room where tourists were still lingering over their desserts and coffee, and was directed towards a banqueting room, which is mostly used for weddings or other family or club parties. Numerous tables at the sides of the room had been laid with spotless tablecloths, and on each was a plate of S shaped biscuits – Burano's speciality – and a number of wine glasses, which would be filled with cool *prosecco* wine to toast de Michelis as soon as his speech was over.

The man introducing him first of all stressed that De Michelis's background, in a 'political' as well as a 'family' sense, was Venetian. While other politicians had recently shown stagnation and an incapacity to act, he had fully proved his dynamism and his political skills in the context of changing East-West relations, and generally in his activities as a parliamentarian and a foreign minister. Previously, De Michelis had conducted the election campaign with great boldness and extravagance, the main slogan was 'not promises but contracts'.[7] On that day, however, he did not put up one of his most flamboyant and exuberant shows, with futuristic projects and promises of glowing profits for all, and his informal talk in Burano was most definitely low-key.

Venice, he said, needed the socialists. He had been a city councilor from 1964 to 1980, and that office was the first springboard of his political career. Now he would have liked to bring his experience of government back to Venice and he was certain that the city would greatly benefit. He and his colleagues in Burano could make 'a unique team'.

> We socialists, have ideas. The *Expo 2000* is one of them, and it may be a great opportunity, but it is not the only idea. Those who oppose it so heartily are just saying – that is all they are able to say – no *Expo* . But it must not be said that I only want to

become Venice's Mayor out of personal vanity or for the sake of a Universal Exhibition, we must also concern ourselves with everyday life and ordinary administration… Burano is certainly a better place than it was in the 1960s, but problems, especially those of housing, health, social services and transport still need attention: I shall certainly feel responsible for the island's day-to-day needs…and, if I do not contribute to their solution and do not live up to my promise, I say to the local team members, whom I know so well, that they are fully entitled to pull me by my shirt tails.

De Michelis's attacks on his rivals were then gone through with business-like efficiency, without the rancour and sarcasm that often seems to characterise political debate.

The Communist Party is in a state of crisis and confusion. The 'Greens' have no capacity to govern. The Christian Democrats are an entirely negative and blindly conservative force. And the place needs a shake-up.

After an attack on the *Lista Civica*, recently formed by one of his ex-socialist colleagues who had abandoned the party, allegedly to pursue the interests of the city, De Michelis ended by inviting electors to keep well in mind all he had told them. The wine was decanted, all present drank his health, and, as the meeting became more informal, a few men approached him and the conversation turned to friendly reminiscence and memories of briefly shared experience were renewed.

De Michelis's short but stylish appearance was followed by that of a Christian Democrat woman candidate. She too was due to speak in the High Street, where a platform had been erected in front of the party's premises. Mixing some dialect expressions with Italian, she began at once with a sharp criticism of the outgoing centre-left council. Their main and most constant failing, she said, was their indecisiveness. Sure to touch on a well known issue, 'they were uncertain on whether to allow the concert of Pink Floyd: that is why it took place.[8] But their weakest area is social welfare'. Appealing to a predictable skepticism about the value of research, she added:

Millions have been spent on *studies*…and what was their conclusion? That it is best for old people to have home helps, or that young people need sporting facilities! Huge sums have been spent in interviews, research projects, consultations, inquiries, but not a penny to help the weakest bands of the city's population…We Christian Democrats assert that *the person* is our first and foremost concern, we must do away with a local government which has shamed the whole city, we must sack all the researchers and all the consultants, put a stop to the proliferation of meetings and discussions. Our leading words are: honesty, uprightness and rigour. If you like those qualities, do not vote for the present centre-left alliance.

Similar themes were those touched on by one of the local Christian Democratic candidates – a veteran of Burano's *consiglio di quartiere,* who added:

201

More room will be made on the council for women, but the most pressing issue is that of ecological problems, and not only because nature is beautiful, but because for Burano protection of the environment is really a condition for living.

Also Cacciari, then De Michelis's principal antagonist in the mayoral election, paid a visit to Burano. While a Christian Democrat Member of Parliament, then representing the provincial government, shouted the last words of his message into a microphone, Cacciari exchanged greetings with some of the old men who habitually spend much time smoking and playing cards in the Communist Party's narrow club room in winter, or dallying by its front door during the hot summer months. His presence seemed to bring back memories of encounters in more militant days.

'Do you remember? I came to hear you in Murano twenty years ago.'
'A long time.'

As the rival speaker closed his talk with a reference to the 'disintegration of Italian Communism; the end of that way of thinking so different from ours', a large loudspeaker above the Communist Party's door started broadcasting the first lines of *Bandiera Rossa*, and some men joined their voices to those of the old record. A local leader introduced Cacciari, emphasising how important it was that he should become the Mayor of Venice: the 1985 election, he said, had brought no improvement in social services, infrastructures or sports facilities, but the Communist Party now showed a strong sense of renewal...and Cacciari's list was particularly attractive to the young.

A woman candidate for the *consiglio di quartiere* made a brief appearance on the speakers' dais to state that it was her hope that there should come about 'a more human society. Women's work in the home should no more be treated as a domestic secret, but as a public virtue'. As in the more intimate and predominantly local meetings of the council, each candidate delivered her or his own message, leaving Cacciari's more extended talk for last. One of them, unconsciously mimicking media advertising, named the qualities of those proposing to lead the island: 'skill, good will, commitment and intelligence' and ended with an invitation, 'Vote with us, vote with intelligence'. Another described all the practical innovations proposed by the Democratic Party of the Left: local crafts, especially lacemaking and boat-building, would be encouraged and subsidised, there would be two flats built for the handicapped, and 'social centres, where old people could pass on their life experiences to the young would be created'. Burano would be directly connected with the hinterland at Ca' Noghera, and parking facilities would be made available at Tre Porti.[9]

Last of all, Cacciari, a true master of exposition, put forward his views. Venice, he said, was a city with extraordinary potentialities. Its complexity was not a limiting constraint, but rather a resource. It offered a wonderful chance for a reorganisation of the whole urban tissue, indeed of its various and diverse urban tissues, from Mestre to the historical centre. His main point was that the Venetian area must not any more be thought of in terms of centre and periphery, but, according to a new and nonhierarchical concept of policentrism. Separating Venice from Mestre, a proposal long debated by 'insularists' (and repeatedly voted against) would have led to development of two peripheries: one industrial and one 'tourist'. By contrast, his idea of policentrism was not, as that envisioned in the 1930s, a hierarchical one in which the centre dominated, but one composed of different, yet equal and complementary units. As he was speaking, Buranelli were clearly – perhaps just temporarily – taken with his ideas. According to that vision, Cacciari continued,

> Burano *is an urban centre*, not a periphery. We do not think of it as a periphery. The lagoon too must be conceived as part of a larger metropolitan area.[10]

Despite De Michelis's extravagant promises, he explained, if it had been decided to hold the *Expo 2000* in Venice, that would have undermined any other plans for the city; it would have destroyed any housing policy, pushing up property prices to the sky; it would have created employment during the years of preparation, but would have left a vacuum and brought the city to a state of collapse once it was over. By increasing tourism and encouraging speculative processes, it would have reinforced destructive tendencies already at work and it would have excluded all other initiatives. His list of candidates, he said, represented a political innovation; they are not only 'morally transparent', but, above all, 'competent, indeed, the most competent, to govern a city like Venice'.[11]

De Michelis's plans, as Cacciari continued, had been conceived entirely within 'a logic of emergency' – it would have led to fragmentary actions and interventions. What was needed instead was '[our] normal, quotidian, and continuous attention to the great patrimony of which we are the heirs.' For a Greater Venice, there was needed a surface Metro, joining Venice-Padua-Treviso. To reorganise the northern lagoon, it was necessary to link Burano with Ca' Noghera. Concerning housing, the need was not only for economical homes, but also homes for people with medium to medium-high incomes. As well as new institutions, the city needed to be given new functions in addition to its traditional ones.

Touching on the age-old problem of antagonism and rivalry between Venice and parts of its hinterland, Cacciari pointed out that the Region, and especially its most dynamic and economically successful cities, such as Vicenza, Padua and Verona, did not want Venice to be the regional capital. The regional government, which was responsible for the environment, had done nothing to stop agricultural pollution. There was also a plan to start a school for restoration in Vicenza (with the support of De Michelis), but would Venice not have been a much better candidate for such an institution?

> Restoration – and not just of art works, but of the city's life in all its aspects – offers us all a great opportunity, if you vote for the list of *Il Ponte* – as I hope, led by me, *with me*, the Mayor of Venice.[12]

Consigli di Quartiere

In various parts of this work I have referred to Buranelli's feelings that they are, in some ways, different from other Venetians, that they have been badly treated, and that, partly due to their isolation, they have been deprived of opportunities to change, to improve their homes, or gain access to desirable employment. Since the 1950s, Burano's economy has greatly improved, but as I found during fieldwork, some of its inhabitants continued to feel that the quality of their lives still left much to be desired. The *chironomidi* crisis, and even more markedly the debates during the encounter with Venice's Mayor which followed it, rapidly became opportunities for the islanders to give expression to all their grievances and long-standing frustrations: compared with other areas, they said, they were badly neglected, their housing and social services were very poor indeed and their transport facilities inadequate.

The venue for such exchanges was that of the *Consiglio di Quartiere* – for Burano an important institution that provides a needed channel for communication with the municipality and for the running of the island's political and administrative life. Indeed, Burano's politics in the 1980s can be viewed as an interesting example of processes of democratisation which, in different ways and different measures, were then taking place throughout the country (cf. Putnam 1993: 17–62).

Briefly, to explain, district councils were brought into existence in the mid-1970s in the context of Italy's general move towards decentralisation, planned since the postwar 1948 Constitution to counteract any potential totalitarian resurgence within the unitary state. In the same spirit it was thought that Italians needed to be radically trained to 'make democracy work' and among the key words of political activists during the 1960s and 1970s was *partecipazione*, participation: everyone should be well informed on planning and economic choices, in order to exercise full political rights. *Partecipazione* therefore was, and is, encouraged – not without a degree of didacticism – on all sides of the political divide, but is particularly congenial to the left, given that absence from political life is viewed as an expression of hopelessness in the face of a superior power structure that people regard as too pervasive or elusive to be influenced or changed, and, therefore, it connotes the kind of passive surrender through which fascism was allowed to suppress Italy's parliamentary democracy in the 1920s.

Another key word, and, of course, related concept is *comunicazione*, communication. Both participation and commuication, it was thought, would be greatly enhanced through the institution of councils with elected representatives, in which views could be aired, needs put forward, plans assessed, and, above all, pressure exercised from the base and claims brought to the attention of the Municipality. Indeed, since their establishment, the councils have held debates on a variety of practical as well as ideological subjects, they have organised cultural activities, attended to individual claims, and, sometimes in conjunction with social workers, they have intervened in cases of hardship, and especially in problems due to housing and evictions.

In her analysis of the relations of a Roman *borgata* with the city's municipality – a situation clearly comparable to that of Venice and Burano – an Italian sociologist observed how self-awareness of the group, which she defines as 'ethnic' – too strong a term, I think, for a district or village in its relations with its neighbours – is a 'collective representation through which a group of people who conceive of themselves as a unit' define their relations with other groups.

> Sharing a common language, territory and traditions, therefore, inspires the group to act in favour of their needs…Sometimes, however, …the 'us versus them' becomes outright 'us against them'. (A. Signorelli, unpublished paper)

Despite vigorous protests and polemical outbursts, however, in the Venetian context a tendency always to seek consensus is very pronounced, but a description of Buranelli as trying, in Signorelli's phrase, 'to remodel their relations' with Venice on a more equal basis than in the past, is nonetheless true to life.[13] For politically motivated members of Burano's council, loyalty to the island and loyalty to their party are often at odds, or are sometimes confused, and at other times balanced against one another.

A realisation that ideologies had become dangerously divisive frequently led to the founding of independent election lists, parties and associations, such as, for example, 'The Young for the Island' or 'Together for Burano, Mazorbo and Torcello' (*I Giovani per l'Isola*, or *Insieme per Burano, Mazorbo e Torcello*). In general, therefore, although Buranelli have a lively interest in national elections and are committed to influencing a choice of government, they find national politics confusing and remote, and the vicissitudes of Italian governments quite adverse to to the efficient running of Burano's and Venice's affairs, while their political battles are always tied in with debates over practical policies and concrete issues.

From my point of view the council's meetings offered a privileged perspective on local sentiments and political attitudes, as well as modes of participation in the management of village life. In Burano, ideas of representation and some degree of local governance were by no means entirely new; on the contrary they had deep roots in the island's civic and associative traditions. Throughout the Middle Ages, like other islands, Burano had an assembly, the *Consiglio della Magnifica Comunità di Burano*. It included forty of the most prosperous men of age above twenty-five; it had the power to elect judges, notaries, a chancellor and heads of neighbourhoods and was represented in Venice by *Gastaldi*. These were substituted by podestà in the thirteenth century, and according to De Biasi, statutes for Burano and Torcello were issued by the Venetian state in 1462 (but cf. Cozzi, above).

Indeed, just because of the remoteness of their island, Buranelli are more conversant with the idea of representing themselves to outsiders than are the inhabitants of the historical centre, where loyalties are more diffuse and districts less clearly demarcated by boundaries. A capacity to represent the island is very much admired, although, as many people remember, the way in which practice was acquired in the past was by putting their case humbly before bureaucrats and people in authority. As we have seen in my chapter on history, memories of a medieval parliament still pervade some of the islanders' patterns of thought and behaviour. Mediation, be it political, social or religious, is sometimes viewed as a necessary step towards improving or restoring the position and fate of the community, as well as that of individuals. In the past petitions for improvements to the island in general, and appeals for help for some particularly needy person or family, were often presented to visiting notables and politicians as they addressed the Buranelli in their public meeting rooms or in the Piazza.[14]

According to peoples' memories of fairly recent events, women who specialised in pleading for mercies or favours, or in vindicating rights and overcoming bureaucratic hurdles, always acquired high status in the eyes of the community – a custom which, as Boissevain points out, bears 'striking similarities [to] the use of intermediaries in the religious field' (1974: 79–80).[15] For example, when, in the 1950s the then President of Italy, De Gasperi, visited the island, at a time when many children suffered from tuberculosis, one of the mothers presented a petition, pleading that the government should finance a parents' trip to the sanitaria in the Alps where the children were hospitalised. Because her request was granted thanks to her skillful eloquence, the incident is one both she and other villagers remember with pride and she was given the nickname of *Avvocata*, 'Advocate' – a word with important religious resonance, since 'Avvocata Nostra' is one of the epithets of the Virgin Mary.[16]

Such faith in social as well as religious mediation, which, Boissevain points out, has been changing in the religious sphere after Vatican Two, is 'deeply rooted in peoples' mentality and not likely to change very rapidly in secular contexts'. Venetian proverbs: 'he/she does not know which saint to turn to', '*non sa a che santo vodarse*'; 'let us hope some saint may come off the wall', '*chissà che qualche santo no se destaca dal muro*'; 'he who has a saint also has a miracle,' *chi g'ha el Santo g'ha anca el miracolo*'; 'without saints, one can't reach Paradise', '*senza santi no se va in paradiso*', and, to emphasise the need to have connections outside the family, 'the saints of the house will never make miracles', '*I santi de casa no fa miracoli*'.

Persons in position of authority (*dottori, professori, tenenti colonnelli*, 'doctors, professors and colonels') are sometimes credited with greater powers than they command in reality. As some Buranelli are ready to recognise and as is shown in the dreams of children and the boasting of adolescents, power and an ability to control reality are mirages which they naively cherish and admire. Accordingly, at times of great poverty and distress, persons whom they regarded as occupying key positions in the hierarchy and who they thought had the capacity to arrange favours and offer help were sometimes appealed to as almost magical saviours.

Power and mediation, important concepts in Buranelli's historical memories, however, had gradually given place to confidence in their own capacities.

By the 1980s, organised political representation had certainly altered the manner and tenor of relations with the outside world, but elements of earlier modes of communication, sometimes characterised by Buranelli's exasperated sense of their own dependence, occasionally coloured their political discourses. For example, when, tired of having their homes invaded by water every time there was a high tide, but forbidden to raise their ground floors by the Special Law for Venice of 1973, and failing to obtain any response from various bodies to which they appealed to have such adverse building regulations waived, some Buranelli decided to turn directly to the then President of Italy, Cossiga. They wrote him a strong letter, which started, 'Help us against the high tides!' At the same time, expressing their anger against the government, and, in ironical parody of the hopelessness of the Venetian municipality and of their own appeals, which they knew would find very little response – and indeed of the very idiom of patronage to which they were resorting in despair – they also wrote to the President of the Italian Parliament, the Minister for Public Works, the Minister for the Environment, the Prefect, the *Pretore*, the President of the Regional Government, the Mayor and, lastly, even to their own district council, that is, in effect, to themselves. In the event, Venice's conservation superintendent, reaffirming the strict conservation principles for which she is well known, instead of granting permission to raise ground floors, proposed as an alternative solution (one sometimes adopted for Venice's palaces), the building of special drainage tubs, which, like narrow moats, would channel the water and keep the houses dry. However, the problem has so far remained unresolved, and the Buranelli predicted that, come November, they would again be paddling in their kitchens and parlours.

As older people remembered, in the past any contacts or confrontations with Venetians were clearly ruled by a division of labour by gender; in general, the task of approaching outsiders, who were automatically assumed to be superiors, on issues concerning individual or family matters, was left to women, and it is they who were expected to plead for help or beg for the attention of some insensitive bureaucrat. This was because women were in any case considered subordinate, so their personal status and pride would not suffer too great an erosion. On the contrary, men, at times when their honour was at risk, were culturally determined, or almost expected, to show temper. They would usually ignore or tacitly support the women's efforts, but they would not ask for help themselves if they could possibly avoid it, as the starting point of encounters in which they were so clearly in an inferior position was always potentially charged with conflict, and expressions of anger and impatience before authority could easily lead them into trouble.

Active membership of political parties was mainly thought of as a masculine prerogative, and, although women were not prevented or discouraged from participating, at the time of my fieldwork the *consiglio di quartiere* was essentially seen as a male preserve. Indeed, in some ways the presence of an active political forum has raised the standing of men in the local community, while at the same time it has altered both the relations between the sexes and those of the islanders with Venice. The institution of a *consiglio di quartiere*, then, added excitement and put a personal note into the business of national and local elections. Councilors were elected as candidates of their political parties, and to some extent the councils have channeled political in-fighting and have also provided an arena in which personal complaints and differences can be aired.

Questions can thus be put before the council, whose members provide advice, if not actual solutions. Family problems, such as those of housing, which in a face-to-face society like Burano's might in any case take on a semi-public nature, are now shifted on to the institutional sphere. According to some of my informants, there is thus a promise that the pressures and influence of personal relations may be checked and decisions rationalised, so that greater weight may be given to rule-bound and impartial administrative practices. Accusations and innuendoes about the fact that people in positions of power acted in self-interest, however, were frequent and sometimes quite vehement, but, as some of Burano's experienced politicians maintain, conflict is itself part of the political process, and deliberations taken through the council are generally viewed as an improvement on personal politics.

Throughout the 1980s, meetings of the council were held in a large room adjacent to the church, once the chapel of St Barbara, where there also took place other indoor public functions such as art exhibitions, political gatherings, lectures, concerts, and so forth.[17] In preparation for public meetings, three tables at the end of the room opposite the entrance door were arranged in a horseshoe pattern for the councilors, while the remaining space was filled with seats for the public. The President of the Council sat at the central table, with representatives of right and left wing parties on either side. Councilors of the opposed parties, Communists and Christian Democrats, thus faced one another and sometimes shouted angry accusations and retorts, or loudly aired their differences of opinion.

Members of the Council often lamented that few Buranelli participated in the meetings, and that people were still new to democratic self-government and open debate, but silence or absence were sometimes justified in terms of timidity and incompetence, not always attributed to lack of interest and cynicism. Meetings are always announced in Venice's daily newspapers, as are those of other Quartieri, on the morning of the day when they are due to take place, and councilors often canvas their neighbours and friends and encourage them to participate. The president of the council in 1988 and 1989 was a young architect – one of a few local graduates who remain in the island, as professional careers are frequently pursued elsewhere and the few who are lucky enough to complete a university education often move to Venice or Mestre.

Because the language of politics is Italian, which is generally acquired in secondary and upper-secondary school – a level which not many older people have achieved – persons in key positions often have to negotiate the choice of language in the conduct of political activities with awareness and sensitivity. During my fieldwork, when the council's president read a motion, or when he delivered a talk of some length and importance, he usually spoke Italian, but when he was addressed in dialect, he answered in dialect, for consistent and exclusive use of Italian would have looked condescending and might have brought ridicule or antagonism. Other councilors chose to speak as they pleased, but the more practised usually alternated between, and rapidly switched from, language to dialect. Some spoke Italian with effort and dropped into the vernacular in discussion. Others spoke Italian without in any way attempting to imitate a standard accent, but their speech was generally closer to Venetian than to 'pure' Buranello, which was considered obsolete and seen as a sign of remoteness and insularity.

Indeed, some of the islanders' capacity to turn their diglossia to advantage due to their ability to go through a whole gamut of linguistic registers and nuances was much admired, and it was indeed a remarkable skill. Thus, for example, when I told my informants that I was interested in their local dialect, and an old man addressed me only in Venetian, he was reprimanded by others for 'talking posh'. If a man is completely incapable of speaking Italian, he may be criticised and considered unfit to represent the island, but 'perfect' Italian speech is regarded as an affectation and a sign of weakness rather than strength. For example, one of the members of the council who speaks at great length in impeccable idiomatic Italian, and even uses the remote past tense, quite outside the range of Venetian speakers who always use the imperfect – and who, even when possessed of a classical education, would regard it as almost disloyal not to do so – often attracts the obloquies of inveterate vernacular speakers. He gets booed and interrupted, he is considered as 'a bit of a sissy', and despite the fact that his political stands are regarded as of the most hawkish, his perfectly voiced complaints that he is 'badly treated just because of his political views' are considered as a show of personal, as well as political, weakness.

Speaking straight dialect, without making any concessions to pressures and without self-consciousness or apology, is considered an act of strength and an assertion of identity. Thus, if a man is prosperous, weighty, experienced and self-assured he may regard forcing himself to speak Italian as compromising his dignity, or as losing a reassuring sense of his own agency and authenticity. On this point too, gender differences are quite marked. For example, one of my women friends told me that she would truly have loved to participate in political debates, but she was painfully conscious of her poor command of Italian. She knew that she had been too strongly conditioned by her school teachers' pressure and negative views of the dialect, but she nonetheless felt that propriety would have required her to speak the language. The viewing of self-possessed and articulate television presenters, she said, had reinforced both her unattainable desire to be 'a little bit more like them' and her self-consciousness. For several other women as well, a fear of linguistic incompetence, added to their timidity about speaking in public, prevailed over a desire to take an active part in politics. They agreed that it was all right for men to speak in dialect, given that 'men are gross anyway', but they assured me that, with few exceptions, women would only contemplate entering politics if they had a good command of Italian. One woman who did serve on a number of committees and who often spoke up at meetings, had learnt both how to muster the attention of an audience, and how to use the dialect to best effect, after years of practice as an actress in the vernacular theatre. In 1988 there was no woman councilor, although there were women on various committees, such as library, parent-teacher associations and so forth, and the running of the council was largely conducted by a highly competent and accomplished woman secretary.

The purpose of the *consiglio di quartiere*, she explained, is both to channel information about policies or events taking place in the city or at regional and national level, and to exercise pressure from Burano on the outside world of bureaucracy and administration, especially at times of political change and elections; councilors sometimes have to negotiate between loyalty to their political party and their commitment to the interests of local voters. The ideal situation, they say, would be one in which a party programme precisely coincided with the interests of the island, but such a situation has never yet occurred.

One of the past council presidents who was described to me as 'the best' representative had shown no deep and unchanging commitment to any party or ideology, and in any dealing with the city authorities had always negotiated in the island's interest. As people noticed without indignation or surprise, but took for granted as a natural and predictable fact, his main concern was with his own advancement, and he eventually became a permanent and well paid employee in the offices of the Regional government. Nevertheless, like other islanders, who having entered politics at the local level have eventually moved on to clerical positions in the Venice town hall or provincial government offices, he was expected to further the interests of Burano and to continue acting as an intermediary in favour of the community, or of individuals seeking jobs, pensions or other benefits. In general, since political ambition and 'putting on airs' are rapidly censured, a way in which to counteract such negative outcomes of a successful career is to show absolute loyalty and availability.

As I noted in my account of meetings, councilors usually spoke in turns, as if to fulfill a ritual obligation and, although the matter in hand may have been entirely uncontroversial, they each felt called upon to show that they had a stake in the discussion. A reason for their eagerness to state their opinions, quite independently of whether or not they could contribute something new, was also to prove that they were vigorous and competent. Therefore, questions and proposals were largely overlapping, sometimes minor points were raised (such as, 'what about the gym?'), or major ones simply restated. Although it was clearly acknowledged that all councilors, as well as people in the audience, have a right to speak, complaints were often made about 'people talking just to show off' or highjacking proposals made by political opponents. 'They only talk to show that they are "more handsome" [*piu' belli*] than others'. Sometimes it was not clear which councilors were to sign a given motion when it reflected the common interest and common will. Indeed, proposals were often duplicated, although the public booed noisily at the repetition.

To mark a new adjustment in relations with Venice, some Buranelli have recently developed a tendency to request that city officials should attend meetings in their island rather than always be under an obligation to go to Venice themselves – one which, they say, puts them in an inferior position. When Venetian officials were present, Buranelli always tried to present a united front and to voice their protests against the state, the region, but above all the municipality, in full unanimity. Councilors were often in agreement on the main objectives to be pursued, but as debates proceeded political differences rapidly came to the surface and antagonisms, whether ideological or personal, were never put aside for very long. Meetings are usually called to discuss specific issues, which are announced beforehand to ensure the participation of persons who may be directly involved. However, discussion does not always revolve on the chosen topics for very long, on the contrary, transitions from subject to subject are often voluble and extemporaneous, as are polemical stances and random interruptions.

Present-day problems, as I have emphasised throughout this work, often invoke the shades of older concerns and anxieties, those too woven in with notions of inside/outside, coloured with nostalgia and with attempts to compare and perhaps revive earlier notions of community, mutual assistance and neighbourliness with the ethics of contemporary caring professionals whom Buranelli regard as lacking in true solidarity and warmth.

On one evening when I decided to sit in on a council meeting, the plan had been to examine the needs of old people, but one of the councilors, rushing in in a state of excitement, explained that something of gravity that had happened the previous night made it imperative that the agenda should urgently be changed to discuss a recent incident to do with the island's pollution. He reported at some length events that he considered extremely worrying and which called for a severe complaint against the personnel of the *Magistrato delle Acque*, once a Venetian office now run by the Italian State, for entirely failing in their duty as watchdogs against pollution.

He had himself volunteered to watch against breaches of rules on pollution control – as he pointed out, at his own expense – and he had actually caught some builders in the act of discharging waste materials excavated at a building site into the lagoon. The *Magistrato*, he maintained, simply does not provide sufficient means or manpower to prevent such flagrant transgressions. Eliciting evident signs of boredom from other councilors he quoted chapter and verse of the articles of law that he had witnessed being broken. Those who suffer the most damage when the lagoon is polluted, he continued, are, of course, the fishermen, 'it is they who should do the policing: it is they who actually *see* what happens all round them, and who have a direct interest in keeping the waters clean'.

It was, in any case, his firm view that silence in the face of abuse was wrong, 'The council should take steps, otherwise we shall be like the famous monkeys'. The problem, then, was how to deal with the *Magistrato delle Acque*, whose duty it was to protect and safeguard the health of the lagoon. In their continual execration of the *Magistrato* Buranelli were at one with many Venetians, and indeed complaints about its inefficiency were an example of comparisons between past and present, the local community and the impersonal and incompetent state. It was, to use Michael Herzfeld's concept, an expression of 'structural nostalgia' that, arguing for the islanders' special relation and intimate knowledge of the lagoon, truly went beyond the community, and encompassed the people and their environment.

Indeed, the very name of that office, itself implying that by Venetian law competent hydraulics engineers were appointed to monitor, and possibly regulate, the natural movements of waters, evokes an almost magical quality. Like other offices of *Ancien Regime* Venice, that magistracy itself was, then, viewed as a symbol of human rationality and of the triumph of order by exemplary legislators who, mustering exceptional engineering talent, could keep the peace over the regime of tides. Criticism of the *Magistrato* , an official of the Italian state, whom people consider a stranger to the Venetian environment, thus shows the full intensity of alarm over the bad state of the lagoon and of a feeling that much had gone wrong after Italian unification.

The president of Burano's council agreed that a letter of complaint should have been written and sent to the *Magistrato* by registered mail. To end, the speaker added that in the matter of pollution control Burano should be a privileged quartiere – and here, of course, an implication of his speech was: pollution control also meant jobs and positions, and, since damage to fishing had affected the livelihood of some of the islanders, they should be the first to have access to those jobs. That aspect, however, was not mentioned explicitly, although it was pointed out that one of the *Magistrato's* employees is a Buranello, and it was hoped that he would encourage the bosses to take on others as well. As his speech had come to an end, various people offered their views from the floor, creating a rowdy noise by their random and impulsive observations. Someone suggested that the *Magistrato* should be sued, but that was considered all too harsh and finally it was agreed that the president of Burano's council should send a message to his office, as had been suggested in the first place.

Despite the proposal that the council should show complete unanimity, on that occasion as on many others, political loyalties and ideological interpretations of events began to take specific contours through the noise; a Communist Party councilor saw the matter in clear-cut Marxist terms. What the *Magistrato* was doing was in effect privatising the whole lagoon! That was all the more wrong, for as we have seen the Magistrate is an official of the state, by definition 'public'.

The speaker, therefore, urged the president of the council to write a *strong* letter. A socialist councilor, who generally takes a broad sweep at a long list of problems, proposed that a motion should be phrased in the following terms: 'Not a lira has been spent on Burano. People want green issues put before building'. Real disagreement, thus, began to crystallise over his particular choice of words – one which could express a fundamental difference in attitudes to the environment. The proposition, obviously very unwelcome to those who favoured development, was expressed in the simple slogan, '*First* pollution clearing, *then* building', *prima disinquinamento, poi cemento*. Expressed in such clear and unequivocal terms, the principle rapidly provoked the Christian Democratic councilors to put up a loud protest:

> the socialists are trying to bring under their political aegis an issue which concerns all of us…what is more, with their campaign 'sloganeering', they seem to be going back to 1968. The motion should be passed unanimously by the whole council.

The socialist exponent agreed to that, but not without a little emendation:

> as the Christian Democratic councilor points out, building and clearing of pollution need not be mutually exclusive; the Christian Democrats would like to have both *disinquinamento* and *cemento*; indeed they think that the two should not be separated at all. But we socialists insist that 'cement' should not be allowed to enter the motion at all.

As on many other occasions, just as all were protesting against factionalism, the noise in the council room had become very loud and voices discordant. In the end, as a man in the audience observed, they were each 'just toeing their party lines'. At the same time, some fundamental questions were asked by members of the audience: 'there is talk of seven thousand millions assigned to the island; surely spending priorities must be decided'. 'What are the state, and the Consorzio Venezia Nuova doing about Burano?' 'God only knows'. And someone commented, 'the times have changed but not the ways!'

Here, as we can see, differences were substantial, and indeed, the debate resembled those conducted in Venice from the 1960s onwards, when industrial development was sometimes demonised by the left-wing, while at the same time trade unionists expressed strong fears that freezing development altogether might have caused serious unemployment. The speakers, then, clearly shared Venetian uncertainties over environmental policies, but indignation about negligence and delay carried stronger and even more personal notes of outrage and of anxiety than those of inhabitants of the historical centre.

> And what about the problem of flooding, of raising the levels of ground floors. It is not only the 'poor'(now rich) fishermen who suffer. It is *all the poor Buranelli* who get flooded out of their homes! One day, when the real deluge will come, they will just forget us all.

213

At that reminder, a note of real doubt began to weigh over the whole assembly; in Venice, engineers were discussing designs for inflatable gates that would entirely block the circulation of sea water into the lagoon. Again, as with pollution, Buranelli felt that they were the most exposed, and therefore the most vulnerable should errors be made. In particular, the idea that such gates would turn the lagoon into a stagnant lake, albeit temporarily, evoked strong fears of infection and illness. The Christian Democrats proposed that the council should put a motion stating that closure of the port mouths without previous cleansing and purification of the lagoon would constitute a serious danger. Here, alas, the islanders knew that they were talking about matters over which neither they nor other Venetians could exercise full control, or feel confident that they had adequate scientific knowledge about.

Keeping issues, and even a degree of anxiety, present and alive is, nonetheless, part of consciousness raising, and a part which the Buranelli ultimately value. Therefore, although the debate clearly concerned major policy questions and physical problems that they had little hope of wholly understanding, they would have liked to bring their experience of lagoon life to bear. As well as major environmental problems, personal needs, conflicts and protests were never far removed from the public arena of the council, and, as informants asserted, a lot more was going on than was visible and audible to outsiders. In that way Burano seemed faithful to its 'face-to-face' tradition, and its council debates were quite different from those which I have witnessed in other parts of Venice.

During the meeting I have described, an old woman had been sitting on one of a row of chairs at the side of the room, huddled in a coarse woolen shawl. From time to time she had attempted to intervene, or uttered something more like a weak sound of protest than a well-articulated phrase. At last, as the voices began to subside, she turned to the president and councilors in exasperation:

> Is it always you, then, speaking and carrying on like so many whores? What about my seventy millions?

Getting very little response, the woman then got up slowly and walked towards the table in a state of visible anger; one or two of the councilors tried, with little success, to placate her.

> Which one of you do I have to give a good beating to?

The woman's house had been destroyed by a fire; as was later explained to me, she might have been put at the top of a housing waiting-list, had it not been found that a heavily padded coat removed by the firemen was in fact full of bank notes.

> But we have explained it all to you, and to your sons as well. It is not us, it is just your Insurance Company, if anyone at all, who should pay for the damage done by the fire... Go home, lady, don't you see how late it is getting.

A Doctor's Duties

Protest, debate and confrontations with Venetian officialdom at Burano's council meetings throughout the 1980s concerned the shortcomings of the island's medical services. Anxieties over health have deep historical roots: Buranelli are well-aware that Torcello's decadence was mainly due to the spread of malaria, while other nearby islands gradually disappeared, due both to environmental factors and to the failing health of their inhabitants who abandoned the heavy task of shoring up their coasts to avoid total submersion. By contrast, Burano, thanks to a favourable conjunction of water currents and the direction of winds, was immune from those ills. A concern with health, however, is also due to more recent problems, mainly the dangerous spread of tuberculosis in the 1930s and 1940s, some instances of Mediterranean anemia and slight disquiet over past inbreeding.

A strong fear of infection, however, also echoes – and responds to – the many charges by medical and teaching personnel from outside the island that some of its inhabitants' ills came from a lack of hygiene and care. Now, as if to reciprocate such allegations (and in keeping with Venetians' fears of contagious diseases brought to them by travelers from distant lands) many islanders affirm that contamination always comes from outside. When, for example, some Venetians confronted with an invasion of backpackers, suggested that Mazzorbo would have been a good place for a camping site, inhabitants of that island and Buranelli together refused outright on the grounds that such tourists would only lead to the spreading of disease, drug addiction and lice. Strangers have to be kept at a safe distance, and, when they do become part of Burano's life, must be carefully monitored.

A great deal of anxiety, fear of contagion and a strong sense of dependence on the medical profession clearly came to the fore when, after repeated and pressing invitations, three senior health service officials came to Burano to hear the islanders' complaints and requests for better medical services. The main problems were first of all stated in clear and unequivocal terms by the President of the district council:

> Burano's health services operate within a space of 35 square metres, in two rooms and a lavatory. The fact that Burano is an island and that, therefore, much medical care has to be administered locally aggravates the situation even more. But hopefully something is just beginning to move: the doctor who now lives in an apartment above the surgery, owned by the commune, has been promised alternative accommodation in Venice by the municipality, and the rooms which are now his living quarters, therefore, will become part of a new medical centre.

With great calmness, the speaker then pleaded with the health authorities to give their attention to the needs of the *Quartiere* and shorten the waiting time. With that he closed his statement, and invited councilors to speak.

The doctor's occupation of a publicly owned apartment was due to the tradition by which a district doctor, or *medico condotto,* was usually given a home, especially when he was posted to some remote and inaccessible village. In this instance, given the shortage of housing, Burano's administration had asked the Institute for Industrial Reconstruction[18] to find the doctor alternative accommodation, and the Institute had undertaken to do so, provided that they did not have to pay for restoration works. The commune, however, would not invest money to restore premises that did not belong to it but remained the property of the state. As a result, as many Buranelli lamented, while the doctor demanded that a new home be restored for him at the public expense and while the district council were trying to evict him legally, five thousand people had been waiting for years for their medical centre. As on other occasions, representatives of different political parties in turn expressed their views and voiced their protests. After the president's introduction, one of the councilors firmly stated,

> relations with the Health Authority are very very bad. On that point all members of the council are unanimous; we may express ourselves differently, but we substantially all agree. We must appeal to the Health Authority not to neglect their responsibilities.

Because, in this case, the views of different councilors were largely overlapping, elaborations of the president's initial statement took place as if through an agreed division of labour or a deliberate effort at mutual respect on the part of people known to be strongly divided on other matters. A socialist councilor, well known for his vehement resentment about the neglect of Burano, added a strong note of protest about the lack of an adequate health centre. He said, a letter had been sent to the Ministry, but 'Buranelli still do not seem to enjoy normal citizens' rights and normal care. If *they* think we are second- or tenth-class citizens, they should clearly say so. Surely conditions are better in Biafra'. The doctor had not fulfilled his duty towards Burano; he had used the rooms for his dwelling but not as consulting rooms.[19]

Another councilor, a member of the Communist Party, then, taking his turn to speak, mentioned that there were many small needs as well as major ones; for example, there was a pressing demand for a trained nurse to take blood samples, for more pediatric assistance and for domestic help for house bound elderly people. Or, as he said rather ominously, 'we shall become a very heavy *quartiere*'.

One of the Christian Democrats again stated that the relations of Burano with the health authority were very bad indeed. Not only were people still waiting for an adequate medical centre, but when experts had come to inspect the pediatric consulting room, they had been horrified: a pediatrician only came to Burano once a week, although, following a protest by a number of young mothers, there had been a promise that he would visit twice a week at the least. 'Ten years work of the council have been completely wasted.' New voices reiterated and reinforced the points already made; two and a half years had gone by since the doctor had been served an eviction order and offered a flat in Venice. Then all restoration works had come to a standstill because the funds had come to an end, and they were still short of thirty million lire (about 12,000 pounds sterling).

A young man, who is himself a nurse at Venice's main hospital, suggested that, once the house was ready and available, plans should be made to allocate spaces for different uses, such as general practice, consultations, specialist treatments and so forth. And what about a dentist? In Burano at that time there were two dentists; neither was fully qualified and both practiced illegally. He also had a lot of complaints to make; the pediatrician was never punctual. There was no ambulance service for urgent needs. With a low tide and a fog it could take a good hour to get to Venice, while in the local consulting rooms there was no emergency equipment, no oxygen, no drugs. But, when this was explained to the Health Authority, their answer was that the doctor had not requested such things. It was high time health officials tried to understand Burano's special situation and time they faced the problem of personnel.

'Buranelli are very humble', the young man said, 'in the way they ask for their needs to be provided for'; they had been patient and prepared to wait, but they were beginning to feel deeply resentful because of the way they had been treated. 'Sometimes they just *have to* get very angry, and then they go and they pull the doctor by the neck, in order to get him out of the house,' as he himself did when his little girl was ill. On that occasion too the meeting ended on a personal note: in a way that would certainly delight those who view the personal as political, the young man gave a vivid example of what he perceived as a grave social problem by the dramatic retelling of an episode in his own life. 'It was at night last September…The thing is history now, but there still remain some unanswered questions. Is it right that a doctor should refuse to pay a house call and diagnose on the telephone?'

One of the communist councilors said that, on that issue, he entirely supported the man's question:

> we must ask for some elucidations on the duties of doctors and the rights of patients. And as for a first aid service, that is a very real need; when two hundred school children made an inquiry on the island's problems, all of them reported that people were very anxious to have a first aid service.

With that, questions were finished and at long last the official was asked to respond. He first of all formally thanked the president and councilors for inviting him to speak. He assured the audience that he had already written a report and presented a motion to the health authorities on the needs of Burano. Unfortunately, the health service was so over-stretched it was almost reaching a point of breakdown. Even the restoration of Venice's main hospital had come to a standstill because of the lack of funds, and problems had piled up in general. Burano, like its southern counterpart Pellestrina, suffered from even greater difficulties, partly because they were so very far from the centre. The solution would have been the creation of new health service districts; but he was not going to make any promises he was not certain to keep. Before the new surgery was ready it could take a whole year. Meanwhile, the local health authority had already assigned new personnel and a medical team was ready to start work in December.

'At the moment, I know, I am the object of your anger. But I understand; I should be surprised if it were not so'. The need for an ambulance, he continued, was a very real one. It might be assigned in the next financial year, but it would not solve all problems. A helicopter would be more the thing. Even that might come true in a not too distant future. There were, of course, objective difficulties, 'You who live in Burano know it even too well; for those who don't live here, it is an uncomfortable island to work in'.

The problem of taking blood samples was very strongly felt, but since there was a legal requirement that that operation should be conducted in a surgery of no less than eighty square metres, that too could not be solved immediately. Fortunately that would soon be available, but, as for a real medical centre, that would require a larger population. It was in any case a matter subject to regional jurisdiction. To answer questions about doctors' duties, he recommended they should read a booklet, *Tuttosanità*. The night doctor, was, in any case, not quite the same as a first aid service.

At this, the man who had raised the question of a doctor's duties again spoke to put forth his view that: 'It is really administrators and technical experts, not politicians, who should take an interest in these problems. Politicians are just passers by; they change' – an idea that echoed tendencies also widespread at a national level and mainly due to a general disenchantment with politics and politicians.

His mind, however, was still on the incident he had referred to before, and which, he repeated, had left many troubling issues unresolved. His determination to discuss only those aspects of it that could be of public concern was undermined by the personal memory which still 'wants to force its way to peoples' attention and consideration'. About the incident, he has written a long letter to the Mayor; he would not read it then, as it all sprang from a personal event, but there still remained a few questions.

1. Is it right that the doctor on duty should refuse to pay a home call and decide on a diagnosis on the telephone? Sometimes he does not even seem to be willing to listen.
2. Is it right that the doctor should undress? If he is called out, he will spend a lot of time putting his/her clothes on.
3. Is it right that persons other than the doctor should spend the night in that very same room, which on the following day we and our children have to attend as patients?[20]

The visiting official, who, by then, hardly managed to conceal his martyred expression, said an answer to the first question was to be found in Law 392, articles 1 and 13: it was for the doctor to decide if he should or should not think a case was urgent.

As the man recalled, in the instance that had left such a painful mark on his memory, when one night the doctor simply refused to go and examine his ailing child, he just went to the *carabinieri*. (As he added rather menacingly, 'I am not very strong...but, in fact, someone stronger...'). The child had fifty pulses, that was not all, she had a nose bleed, and she had a very high temperature; imagine how a father would feel. Luckily it was not serious. But, after the incident, the little girl did not leave the house for two days for fear of meeting the doctor in the street, or, worse still, the *carabinieri*. Surely doctors should have some sensitivity about the kind of place they work in. They should make some effort to adapt to the social environment. 'If some old lady comes to have her blood pressure measured every day because she feels lonely, that too is a problem'.

As the president of the Health Authority recalled, the incident described by the young man had been fully reported in the Venice newspaper, and he was certainly left in some apprehension when he had read about it. It was the second Venetian official who again answered the man's main question, in the hope of bringing the meeting to a close; it was up to the doctor on duty to decide which cases he thought demanded a house call. He was obliged to keep a record of every call and of every answer he gave in his register. If he failed to intervene in an emergency, that was a very grave offence. As for the consulting room, it was usual that it should have a sofa. The presence of another person was not normally authorised or even contemplated, but whether the doctor did or did not take off his clothes and get into a dressing gown really was irrelevant.[21]

Here we see how the speaker, invoking values of solidarity, compassion, and propriety, which are often thought to have been characteristic of the island in the past, would have liked to see such values perpetuated by persons in the caring professions, in the context of a health service organised from the city. However, in as far as such medical services were run on 'modern' lines, islanders in general regarded them as coldly impersonal and heartless.

The *consiglio di quartiere* meetings, then, provided the setting for the reenactment of the age-old contrast between the island-village and the Venetian administration, the state and the social services. Nevertheless, as well as a forum for protest and criticism it provided opportunities for advancing legitimate demands and for seeking long-needed answers to the community's needs.

Notes

1. Newspaper accounts were by no means detached and objective; in this instance the reporter becames an active participant rather than an impartial observer in the debate, and contributed to precipitating hostility.
2. The Mayor recalled that he had also been concerned with pollution as President of Italia Nostra and had created a Committee which intervened against delays in the implementation of the Special Law for Venice.

3. Buranelli readily identify with, and feel compassion for, all the disadvantaged and poor in the city – or the world, but most of them would reject a view of their society as classless.

4. A *qualunquista* was a member of the *Party of the Average Man,* founded after the Second World War to represent mainly Roman and southern white-collar workers fearful of communism and nostalgic for the fascist past.

5. The Liga Veneta put up candidates at the European elections of 1979. It had a candidate elected to each House of the Italian Parliament at the 1983 General Elections.

6. Vague threats to separate from the Venice commune were often, but not always, vacuous. The Cavallino *quartiere* has recently decided to constitute itself as a separate commune – a decision promptly regretted.

7. De Michelis had gone about tendering contracts that he asked various people to sign. When it emerged that not only businessmen, but also public officials had underwritten such contracts, De Michelis was attacked in the press and threatened with a court case for unconstitutional behaviour.

8. A rock concert that attracted excessive numbers of people and indirectly caused damage to the city. The Council had permitted the concert mostly out of inertia.

9. Parking space was more costly than in Venice's terminal, Piazzale Roma.

10. Cacciari's approach was undoubtedly an intellectualist one. While other politicians accused rivals of 'not-doing', or 'wrong-doing', his main criticism was most frequently 'they don't understand'.

11. Members of the Gramsci Foundation Study Group, and Cacciari's colleagues at Venice's Architectural School, have studied the city's planning for years.

12. The election result showed little change in Burano's political orientation: the Christian democrats (DC) won 38.5% of votes, followed by the communists (PCI), with 21.4% and the Democratic Party of the Left (PDS) with 12.3%, while the Green Party, more successful than in previous elections, won 8.3% of votes.

13. Paradoxically – but only apparently so – conflict and protest tend to confirm rather than undermine the closeness and interdependence of the parties concerned. In family feuds, as well as political debate, a search for settlement of disputes leads to very vocal expressions of disagreement, but some compromise solutions must be reached in the end. For example, in quarrels between siblings, especially when they are tied in common boat ownership, and while parents are alive, a permanent breakdown of relations is most unlikely. The most dramatic shows of hostility and fierce threats, according to some informants, are usually the prelude to reconciliation.

14. When, in 1812, Napoleon visited the island, a Venetian woman appealed to him to found a hospital for 'our poor rickety children'. Visits from people who came to buy or commission lace were innumerable. During my fieldwork, a Christian Democrat Member of Parliament, Tina Anselmi, came to address a political meeting. One woman appealed to her to see that prescription charges should be waived. Another pleaded that she should arrange for her son to be exonerated from army service, but she was brusquely interrupted by people who claimed she had no right to appeal to a Christian Democrat, because she was herself a socialist and should turn to a representative of her own party. Among recent visitors was Prime Minister Andreotti in September 1989.

15. As Boissevain writes (1974: 79–80), 'There are striking similarities between the use of intermediaries in the religious field and the brokerage and patron-client relations which are particularly strong in Catholic countries. The importance of intermediaries, especially in the political field, is summed up neatly in the proverb often quoted by Sicilian and Maltese: "You can't get to Heaven without the help of saints', for political patrons in both cultures are referred to as saints".

16. When infant mortality was high, baby girls were sometimes named Avvocata, as it was hoped that, if they were to die, they would plead with God to remit the sins of their parents.

17. Meetings are now held in the chapel of a long-abandoned convent, the Cappuccine, charmingly restored and used for secular functions.

18. I.R.I. was founded in 1933 by the fascist government to take over much industrial development and to protect medium and small investors during the depression. After the war, IRI was regarded by the left as a legacy of the corporate state, but since the late 1950s and 1960s it has remained the largest economic agency, managing Alitalia, the Telephone Company and some electrical industries, as well as industrial developments in the South.

19. Previously, someone had complained to me that the doctor would not even open the door to patients but hand out prescriptions from the window. However, some people sided with him and, when it was proposed that he should be moved out by force, their response was: 'not on our bodies!'

20. As in many other instances, questions are phrased in terms of propriety.

21. The offending night-duty doctor was a young woman: her sex may have aggravated the sense of indignation and pollution fears, but that aspect was never openly mentioned.

Figure 8.1 Burano's representatives at Venice's municipal palace. 'Comune egoista Give us a help: sport means good health. Give us back our football pitch!'

CONCLUSIONS

In this book I have attempted to describe Buranelli's lives as well as their attitudes to history and their efforts to improve their island's environment and living conditions. Among the theoretical problems that naturally arose both during fieldwork and in the process or writing and reflecting upon my field notes, were questions about the formation and continuity of Burano's collective identity, the usefulness of kinship analyses in complex bilateral societies and the validity of an analytical approach based on 'honour and shame' – a central topic in the study of southern European societies since the 1950s.

An answer to my first question 'on what bases and by what processes do Buranelli define themselves as a distinct community?' partly lies in their relations with persons and institutions outside their island and, in particular, with inhabitants of other parts of Venice. In the 1980s, when I conducted my fieldwork, interaction frequently focused on lively debates about environmental problems in the lagoon. Discussion of proposals for putting repair to large-scale ecological damage thus provided a starting point for observing relations between Buranelli and the Venetian administration of which their island was a highly critical *quartiere*. Not every part of the Venice commune experienced such problems and contemplated their solutions in the same ways. Inhabitants of inland areas, for example, were not as adverse to industrial development as were most inhabitants of insular Venice. In Burano concern focused mainly on the pollution of the lagoon, the poor state of housing, the inadequacies of transport and, above all, of health and social services. Such issues, iterated at political meetings and very often explained to me with a hope (alas, quite unfounded) that I might be able to lend them support, were only one of the bases for Buranelli's sense of their separateness and distinctiveness. Their sense of identity did not derive only from their relations with those they classified as their 'others', but was also based on their deeply felt closeness to their lagoon environment, which informed their cognitive orientation, their language and its metaphors, and posed a constant challenge to their skills, since from it they had traditionally drawn their livelihood.

A need to affirm their continuity in a context of change led Buranelli to reflect upon their past and record memories that would contribute to their individual and collective self-images. Details of language, such as words now considered obsolete, cooking recipes and even customs still practiced and well-known to all, but which they feared I might consider outmoded, were always reported in the past tense or attributed to the old (*i nostri veci*). However, this did not show a disjunction between past and present, but, on the contrary, it showed that past and present were very closely woven together. As for inhabitants of the Val di Fassa described by Cesare Poppi, 'the very fact that they constantly talk about, compare and contrast *chis egn*, (those years, the past) with *inchecondi* (nowadays, the present) in dichotomous terms indicates the degree of closeness – if not identity – with the past that shapes the actors' own perception of themselves' (1992:119).

Indeed, although most Buranelli were eager to participate fully in the benefits of modernity, they were nonetheless unsure and somewhat divided about the extent to which they were, and preferred to be, a 'traditional' community and the extent to which they wished to think of themselves as fully part of the modern, technically advanced and materially desirable, world. In this context 'modern' and 'traditional' emerged as abstractions, both overlaid with affect and ideology. The Buranelli's ambivalence about modernity was also due to their uncertain stand towards the world outside their island. In other words, against the arrogance, or what they perceived as arrogance, and lack of environmental sensitivity of outsiders, they upheld a sense of moral superiority, authenticity and good neighbourliness. Nostalgia for a past that they described as characterised by solidarity and human warmth, therefore, coexisted with sharp competition in business at an individual level and with collective claims for the advantages of modern medicine, technology and transport. However, while as a community Buranelli eagerly sought any subsidies that would have helped improve their environment and social services, for individuals or families to seek help from the state would have been like an admission that they had abandoned the values of self-reliance and pride, and their determination to improve their circumstances through hard work, skill and enterprise.

Buranelli's reconstructions of their history is an interesting example of the weaving together of personal narratives, shared memories and legends handed down through oral tradition, with individual fantasies and imaginative folk etymologies, and with historians' reports of past events – those too drawn from 'common knowledge and hearsay' (Coronelli 1696: 33–34). Through all of these, Buranelli created their own meta-history, focusing on themes such as innocence, humility, divine providence, but also cunning, enterprise, skill, and above all family love and solidarity.

Contrary to an idea widespread throughout the 1970s that, as anthropologists turned to the study of complex societies, kinship analyses would prove irrelevant (but see Just 2000: 5–7), I found that understanding of Burano's kinship structure and kinship idioms provided a great deal of insight into their sense of community, while family sentiments and values were often projected onto the sphere of religion and piety. Buranelli's vision of their community as one rendered solidary and compact thanks to its multiple kinship links was also an oppositional one – that is, one by which they clearly differentiated their island from the city and the outside world. In descriptions of their strong attachment to family, they emphasised their need to rely on their kin for mutual help, so that branches of a family, especially those established outside the island, were viewed as necessary paths for those wishing to integrate in a world beyond and outside Burano, which they considered closed and hard to penetrate.

Comparison between forms of residence and economic organisation in Burano's fishing community and Tre Porti's extended families of agriculturalists showed important differences in attitudes to authority and, above all, gender relations (Loizos and Papataxiarchies 1991: 3–25). However, an increasing tendency for members of both communities to reside neolocally was gradually reducing past differences and forms of domestic organisation.

Although Burano is commonly referred to as the 'island of lace', in contrast to Murano, 'the island of glass', before conducting fieldwork I had not anticipated that lace and lacemaking were such important symbolic as well economic factors in the islanders' lives and self images. As a quintessentially feminine craft, associated with ideas of sexual and domestic virtue, lacemaking could usefully be analysed through an 'honour and shame' perspective, indeed, it naturally led me to give greater space to 'shame' – to my mind the more challenging of the two terms, but overshadowed in earlier ethnographies in which greater attention was given to 'honour' as a mainly male prerogative.

A focus on gender, emphasis on subjectivity, foregrounding of individuals and collection of extensive life-histories, as well as the history of lace and its making, all proved very revealing of the moral ideas and prejudices that informed women's work in the past. In particular they showed that a view of women as liable to sexual transgression and 'shame' was always prominent in the discourses of its organisers. Patriarchal traditions and notions of honour and shame converged with religious ideas in a dominating concern with sexual purity and restraint.

A question, that naturally arises in reflecting upon Burano's history is: did the Catholic Church support or did it hinder the freedom and emancipation of women? A positive answer can be summed up in Jane and Peter Schneider's view that:

> in their confrontation with lineal descent groups both the Church and Islam *sought to buttress the position of women against that of men* [my italics] They granted women rights and status denied them by their families [while]...familism and the code of honour defended the family against the hegemony of Church and State. (1976: 96 cited in J. Davis 1984: 32)

Eric Wolf (1984: 7) similarly maintains that the Church aimed to 'lessen the claims of kinship over the conjugal pair', and was 'involved in spreading contract law' (instead of laws based on 'status'), thus naturally lessening patriarchal authority. However, evidence from Burano goes against such positive views of the Church's influence, because, while the Church may have undermined fathers' authority over *sons* to oppose powerful lineages, it certainly did not oppose their control over *daughters*. On the contrary, throughout its history, the Church increasingly advocated virginity and chastity to such an extent that a woman's chances to take up salaried work outside the island were gravely restricted, while sexual freedom was viewed as inevitably leading to shamelessness and sin.

There is some truth in the Schneiders' affirmation that the Church 'sought to buttress the position of women' in as far as the capacity to work for cash, and the fact of being early socialised in the Lace School helped to make many of Burano's women more eloquent and assertive than women in other areas, but words that appeared most frequently in the lacemakers' recollections were 'exploitation', 'sacrifice' and even 'enslavement'. Indeed, convents that usually mirrored the social and economic differences of the world outside, were potentially empowering for 'daughters of the nobility' (Coronelli 1696: 33), but not so, my informants maintained, for their poorer sisters, for whom life in, or at the margins of, convents was inevitably one of renunciation and hard work.

The history of women's work thus shows how in the changing political conditions of Italy in the nineteenth century, relations between Church, state and family underwent considerable changes in doctrinal emphasis and orientation. In particular the Church addressed itself differently to gendered aspects of honour: it was at one with the state in as far as it aimed to diffuse bellicose lineage loyalties and feuds, but in its attitude to sexuality and marriage it was at one with patriarchal attitudes in supporting full control and authority over women. For example, we find that Church and state enjoyed a period of unprecedented harmony after the 1929 Concordat in the early years of fascist rule. It is no coincidence that expulsions from the Lace School for 'immorality' go back to those years. However, the Church and secular forces were again at odds from the 1940s onwards, and particularly in the 1970s, when in response to pressure from the left and from women, the state passed legislation on divorce and abortion that is in opposition both to the Church and to traditional familial outlooks.

The uncertain feelings of some of Burano's women about lacemaking were clearly in evidence when, in the 1980s, a plan was made to open a museum of lace and offer lacemaking classes in the old School premises. The contrasting prospects of museification, i.e., the relegation of lace to the past which to them appeared like a definitive closure, and of revival, which brought back memories of exploitation, gave rise to contradictory views and emotions. Resentments about the past, and a fear that the renewed presence of the School might have interfered with their independent trade, then were clearly part of an ongoing political discourse.

Several of my male friends in Burano assured me that they too retrospectively resented the pressures that had been exercised on their women by paternalistic outsiders; emphasis on their need for education and control cast a negative light on the community as a whole. Therefore, while discipline based on convent rules may have coincided with men's desire for control, it ultimately brought humiliation on the men as well, by keeping their preoccupation with female sexuality and their fear of dishonour ever alive.

In 1970s and 1980s Burano, therefore, the history of lacemaking acquired a marked political dimension, but political activity was mainly a male preserve, exercised in the male spheres of the main street and the *piazza* or the cafés, party premises, and meetings of the council. In discussions about measures to prevent flooding and pollution, Buranelli often felt challenged in their sense of honour and their pride in their intimate knowledge of the lagoon. They thought their own folk-knowledge and experience should have been taken account of, in preference to much scientific work (Menez 2000: 33–41), while memories of past isolation and neglect were recalled to support their negotiations for new housing, and improvements to their health and social services.

Analysis of political interaction between Venice and Burano also showed that, although some relationships bore the characteristics of clientship, in as far as they were openly associated with exchanges of favours or potential openings to advantageous relations and contacts, they were usually based on equality and mutual freedom of the parties involved. Like 'kinship', 'honour' and other terms that have become quasi-technical anthropological key-words, 'patron-client relations' would need rethinking, because it covers many different kinds of connections, links and exchanges (cf. Pardo 1996:12 & 57–60).

Integration of Burano's community in the larger structure of the Venice commune in the 1970s and 1980s was conducted through different idioms from those of the past, as awareness of social and political rights had largely displaced older discourses characterised by deference and social distance. Changes in administrative and political power, especially in the last thirty years – with the formation of regional governments and the development of European unity – have been fast and in many ways disorienting. For their part Buranelli have been very dynamic in participating in the global economy: part of the lace now sold in Burano's shops is made to their orders in the Far East. Nevertheless, while determined to integrate into the larger political entities and global economies, and to negotiate the local community's needs and formulate its demands of the state, regional, party and municipal bureaucracies, Burano's representatives at the same time firmly asserted their island's social identity.

A new vision of a 'polycentric Venice', advocated by its Mayor, Cacciari, is certainly appealing: indeed, with improved transportation and an increasing use of information technology, perception and experience of life in the island periphery may entirely change.

APPENDIX 1

❧❧❧

The Venetian Territory and its Population

Over the last two hundred years, administrative units within the Venice commune have undergone several major changes. Before Napoleon's invasion in 1797 and his introduction of new subdivisions, Venice and its surrounding territories were described as a Duchy, or *Dogado*. Its borders never corresponded exactly with the physical borders of the lagoon, but varied according to the political and military expansion of Venetians into nearby areas. Outlying villages were commonly referred to as *contrae* (Italian, *contrade*) which literally means 'settlements at the sides of roads', that is, hamlets or boroughs, generally equivalent to parishes.

At the time when Venice was joined to Italy in 1866, its area had contracted and its boundaries more or less coincided with the confines of the lagoon. The present territorial extension, which has sometimes led to comparisons of the Venice municipality and of its administrative problems with Greater London, was gradually brought about through subsequent aggregations of neighbouring areas. Lido and Malamocco were joined to Venice in 1883, Marghera in 1917, followed by Pellestrina in 1923, Murano, Burano, and Ca' Vio in 1924, and Chirignago, Zelarino, Mestre, Malcontenta and Favaro Veneto in 1926. The word commune, however, is sometimes used ambiguously, as it still designates smaller units, such as villages or groups of villages, which were at some time autonomous, or semi-autonomous. Indeed, some localities retain partially separate administrative identities, and they have recently been constituted as districts, *quartieri*, with their own elected councils, which, in the outlying villages, still meet in their ancient municipal halls.

Changes, both to the boundaries and to the internal subdivisions of the commune make diachronic comparisons very laborious. For example, in the 1971 statistics the historical centre, terra firma and estuary, are subdivided into *sestieri* (according to the traditional division of the city into six parts) and *frazioni amministrative*, or 'localities'. In the 1981 census, however, the entire territory is divided into 18 *quartieri*. Some internal divisions in the historical centre remain the same as those of old *sestieri*, but others are somewhat changed, in particular, *quartiere 1* includes San Marco and the two old *sestieri* of Castello and Sant'Elena. Cannaregio remains *quartiere 2,* Dorsoduro (3) includes San Polo and Santa Croce, and Giudecca (4) includes Saccafisola. Comparisons of population figures before and after the institution of *quartieri* in 1976 require rather complex topographical, as well as arithmetical, operations. Burano, however, remains one *quartiere,* and its inclusion of Mazorbo and Torcello follows a long-standing tradition, and conforms to the pattern of its classification as a 'locality' in 1971. Similarly, the three larger zones (historical centre, terra firma, and estuary) remain mainly unchanged.

It is interesting to note that the numbers of residents in the historical centre is now approximately equal to the figure for 1866, which was then thought to have reached an all time low. The population of Burano, as well as of other 'traditional' areas such as Pellestrina, Mazorbo and Sant' Erasmo, generally follow patterns similar to those of the historical centre. In either case, development of new housing is almost impossible, partly due to the limitations of space and partly to problems of conservation. The areas where the population decrease has been the most marked are those of Venice and the estuary islands. By contrast, population figures have followed national trends in the industrial hinterland areas of Mestre-Marghera, and also in Cavallino and Lido where trade, tourism, and other tertiary sectors have developed significantly, and where new neighbourhoods have been built with no constraints due to conservation laws.

APPENDIX 2

꧁ꕤ꧂

Law 16 April 1973, n.171. Interventions for the safeguard of Venice

The Chamber of Deputies and the Senate of the Republic have approved.
The President of the Republic promulgates the following law: Title I, Article 1.

'The safeguard of Venice and of the lagoon is declared a problem of pre-eminent national concern…The State, the Region and the Local Authority, each within the ambit of its competence, will cooperate for the achievement of the aims described.'

As is specified in subsequent articles of the law, 'the task of determining the territory to which the law itself would apply was assigned to the Regional authority; futher indication as to future development, measures for the protection of the natural as well as artistic environment, would be elaborated by six cabinet ministers, with the president of the regional junta, the president of Venice's provincial administration, the Mayors of Venice and Chioggia and two representatives of other communes, and published by the central government.'

More attention is given to conservation than to Venetians' economic future, while regional and provincial authorities figure very prominently in comparison with representatives of the historical centre. Regulations are very severe, and any restoration or change in the lagoon's green belt, the islands of Pellestrina, Lido and Sant'Erasmo is strictly forbidden.

Article 13 concerns regulations on building and restoration in the city. Responsibility is left to the commune; decisions must be based on norms due to be elaborated by a parliamentary commission formed by ten senators and ten deputies nominated by the respective assemblies and by the Region. Its main criterion will be: interventions shall be in accordance with plans adopted by the commune and shall be supervised by the Superintendent of Monuments. Exception made for monumental buildings for which conservative restoration is always approved, interventions will be carried out on discrete areas with a unitary character, and in such a way as to maintain the structural and typological characters of the buildings included…In the field of minor buildings, the state offers a repayment of 40% for restoration expenses incurred by private owners. (Conditions placed for such repayments, however, are of such bureaucratic complexity that few people have benefitted so far).

Various sums are allocated by the government: for the years 1973 to 1977, respectively 25, 60, 90, 85 and 40 thousand million lire. However, delays in planning and in reaching agreement at the local level on works proposed within the given deadlines make it very difficult for Venetians to benefit from government funds.

APPENDIX 3

❦

Census

Census figures for 1991 show considerable change in Burano's population from 1981. The number of residents has gone down by 17.5 %, from 5,208 to 4,299, with a loss of 909 individuals. As for the whole of Venice, and especially the historical centre, there has been a noticeable ageing of the population: while in 1981 people over 60 were 773 (15%); in 1991 their numbers had risen to 970 (22%); 58% women and 42% men.

By contrast, the number of children of ages 0 to 15 has gone down by 52.5 %, from 1,143 to 543. For ages 15 to 59, the population has gone down by 15.4 %, from 3,292 to 2,786.

Table A3.1 Population of Burano according to the 1991 census.

	0 to 14	15 to 59	59 and over	Total
Females	275	1,329	560	2,164
Males	268	1,457	410	2,135
Total	543	2,786	970	4,299

The most noticeable change in employment statistics is a marked increase in the numbers of students, from 139, 2.7% of the total population in 1981, to 232, 5.4% of the total population in 1991. The main change in occupational structure is a smaller number of people engaged in 'fishing, hunting and agriculture' from 166 to 136 and an increase in the number of persons employed in the building industry, from 84 to 167, as well as a slight increase in the number of women employed in trade, from 156 to 164.

There is also an increase in the numbers of workers, however, numbers in the category 'ordinary worker' have been overtaken by *operaio specializzato*, skilled worker, especially for women, of whom those classified as skilled have risen from 43 in 1981 to 103 in the 1991 census.

Educational standards are somewhat lower than in other parts of Venice, especially in the historical centre,

Table A3.2 Educational qualifications for ages over 6 in Burano, 1991.

	Males	%	Females	%	Total	%
University	11	0.3	3	0.01	14	0.3
Diploma	168	4.1	114	2.8	282	6.9
Lower Middle School	791	19.2	616	15	1,407	34.2
Elementary School	757	18.3	898	21.7	1,655	40.0
Can read and write	250	6	344	8.3	594	14.3
Illiterate	73	1.8	103	2.5	176	4.3
Other			1		1	
Total	2,050	49.7	2,079	50.3	4,129	100.0

The number of female university graduates has risen from one to three, but male graduates have gone down from 18 to 11 – most probably due to the fact that male graduates are among those who have left the island. It is of note that the numbers of persons with diplomas and secondary school education, respectively 3.3% and 20% in 1981, have gone up significantly to 6.8% and 35% in 1991, while persons with elementary school certificates, 46%, as well as those defined as illiterate, 6%, or semi-literate, 24% in 1981, have gone down to 40%, 4% and 14.5% respectively (all of them over age 55). Burano's educational standards are nonetheless lower than those of Venice's historical centre, where 9% of people have university degrees and 23% have diplomas, and of the commune, where university graduates are 5.5% and persons with diplomas are 22.5%.

Due to the exodus of population, as well as the construction of new homes in Mazorbo, the pressure on housing has greatly diminished, thus, while according to the 1981 census almost 14% of families were forced to live in shared accomodation, in 1991 families in excess of the number of houses were only 7, or 0.3% of the number of families. 77% of houses were owner-occupied, and 20% rented, and the remaining 3% 'under some other title'. Housing, therefore, is no longer a problem, while the islanders are concerned about their diminishing numbers and the low birth-rate.

BIBLIOGRAPHY

Allen, N. J., 'The field and the desk: choices and linkages', in P. Dresch, W James and D. Parkin eds., *Anthropologists in a Wider World*. Oxford 2000

Anon., *La Perfettione del Disegnodi Varie Sorte di Recami e di Cucire*. Venezia 1576

Appadurai, A., 'The past as a scarse resource'. *Man* 16, pp. 201–219. 1981

Ardener, E., *Social Anthropology and Language* (ASA Monograph 10). London 1971

————, *The Voice of Prophecy and Other Essays*, Ed. M. Chapman. Oxford 1989

Ardener, S., *Defining Females: the Nature of Women in Society*. London 1978

Aretino, P., *Vita di Maria Vergine*. Venezia 1545

Argyrou, V., *Tradition and Modernity in the Mediterranean*. Cambridge 1996

Barbagli, M., *Famiglia e Mutamento Sociale*. Bologna 1977

Barbaro, R., *Burano 1903–1964*. Milano 1989

Barizza, S., *Il Comune di Venezia, 1806–1946. L'Istituzione, il Territorio. Guida-inventario dell'Archivio Municipale*. Venezia 1987

Barthes, R., 'Historical Discourse'. in Frederic, L. *Structuralism, a Reader*. Trans. P. Wexler. London 1970

Battaglini,T., *Torcello Antica e Moderna*. Venezia 1871

Battisti, C. and Alessio, G., *Dizionario etimologico italiano*. Firenze 1950–7

Bellavitis, G. and Romanelli, G., *Venezia*. Bari 1985

Bellavitis, G., *Difesa di Venezia*. Venezia 1969

Bellemo, E., 'Il Folclorismo Peschereccio nei centri marittimi della laguna di Venezia', in G. Magrini ed. *La laguna di Venezia.*, vol III, part 6, book 9, pp. 261–352. Venezia 1940

Bellodi Casanova, L., V. Cargasacchi, D. Davanzo Poli, M. Gambier, M Levi Morenos and A. Mottola Molfino, *La Scuola dei Merletti di Burano*. Venezia 1981

Benveniste, E., *Le Vocabulaire des Institution Indo-Europeenes*. Paris 1969

————, 'Categories of Thought and Language', in *Problems in General Linguistics* (1966), trans. M. E. Meek. Miami 1971

Berlin, I., *Concepts ad Categories. Philosophical Essays*. London 1978

Binaghi Olivari, M.T., *I Pizzi: Moda e Simbolo*. Milano 1977

Bloch, M., *Marxism and Anthropology*. Oxford 1983

Blok, A., *The Mafia of a Sicilian Village*. Oxford 1974

————, 'The Peasant and the Brigand: Social Banditry Reconsidered', *Studies in Society and History*, 14, 4: 494–503, 1972

————, 'Rams and Billy-goats: a Key to the Mediterranean Code of Honour.' *Man* 16: 427–40, 1981

Boissevain, J. and Friedl, J. eds., '*Beyond the Community. Social Process in Europe*'. The Hague 1975

Bonaccorsi, G., ed., 'Protovangelo di Giacomo', *Apocrifi*. Firenze 1948

Bonesso, G. F., 'Granchi in laguna. La produzione delle *moéche* a Burano'. *La Ricerca Folklorica*, N. 42. pp. 9–26. Brescia: Ottobre 2001

Bordone, B., *Libro nel quale si ragiona de tutti l'isole del mondo*. Venice (1528) 1565

Boschini, M., *La Carta del Navegar Pittoresco*. Venezia 1660

Bouwsma, W.J., *The career and thought of Guillome Postel*. Los Angeles 1957

Brandes, S., 'Reflections on honor and shame in the Mediterranean', in Gilmore, D., ed., *Honor and shame and the unity of the Mediterranean*. Washington 1987

Braudel, F., 'La Vita Economica di Venezia nel Secolo XV', *La Civiltá Veneziana del Rinascimento*. Venezia 1958

————, The Mediterranean and the Mediterranean World in the Age of Philip II (2nd ed. 1966). London 1973

Brooke, M.L. and Simeon, M., *The History of Lace*. London 1979

Brown, H.F., *Life on the Lagoons*, London 1909

Brunello, F., *Arti e Mestieri a Venezia nel Medioevo e nel Rinascimento*. Vicenza 1981

Brusatin, M., *Venezia nel Settecento. Stato, Architettura e Territorio*. Torino 1980

Burman, S., *Fit Work for Women*. London 1979

Campbell, J.K., *Honour, Family and Patronage: A Study of Institutions and Moral Values in a Greek Mountain Community*. Oxford 1964

Carocci, G., *Storia dall'Unitá a Oggi*. 1975

Castets, F., *Dante Alighieri 1261–1321. Spurious and doubtful works. Il Fiore*. Paris 1881

Ceresole, V., *Les Origines de la Dentelle de Venise a l'Ecole du Point de Burano a l'Exposition de Paris*. Paris 1878

Cessi, R., *Documenti Relativi alla Storia di Venezia anteriori al Mille. Testi e Documenti di Storia e di Letteratura Latina Medioevale*. I, Padova 1942

————, *Storia della Repubblica di Venezia*. Messina 1944

Chambers's Twentieth Century Dictionary. Glasgow 1949

Chapman, M., *The Gaelic Vision in Scottish Culture*. London 1978

Chiancone, M., *Relazione al Commissario Straordinario per il Comune di Venezia sull'Annessione dei Comuni di Murano e Burano a Venezia*. Venezia 1924

Chinello, C., *Porto Marghera. 1902–1926. Alle Origini del 'Problema di Venezia'*. Venezia 1979

Collier, J. F. and Yanagisako, S.J., 'Toward a Unified Analysis of Gender and Kinship.', in *Gender and Kinship. Essays toward a Unified Analysis*. Stanford 1987

Chretien de Troyes, or Yvain. *The knight with the Lion*, trans L.J. Gardiner. London 1912

Clark, M., *Modern Italy. 1871–1982*. London 1986

Cohen, A.P., 'The same – but different!'; the allocation of identity in Whalsay, Shetland'. *Sociological Review* XXIV: 449–459, 1978

Comune di Venezia, *Censimento della Popolazione 10 Febbraio 1901*.

————, *Venezia, Problemi e Prospettive. I Servizi Giornalistici*. 1967

————, *Piano particolareggiato del centro storico di Burano* (not dated)

————, *Piano Programma 1977/1980*

————, *Censimento e Serie Storica della Popolazione* (1950–1980). 1981

————, *Il Censimento a Venezia. Dati Per Quartiere.*. 1981

————, *Informazione Statistiche. Annuario.* 1988

————, *Informazioni Statistiche* 1989

Conton, L., *Torcello, il suo Estuario e i Suoi Monumenti.* Venezia 1749

Corner, F., *Ecclesiae Venetae.* Venezia 1749

Coronelli, V. M., *Isolario dell'Atlante Veneto. Cosmografia della Serenissima Republica Veneta.* Venezia 1696

Correr, G., *Venezia e le Sue Lagune.* Venezia 1847

Cortelazzo, M., *Guida ai Dialetti Veneti*, Padova 1979–1993

————, *Parole Venete.* Vicenza 1994

Cozzi, G., *Repubblica di Venezia e stati italiani: politica e giustizia dal secolo XVI al secolo XVIII.* Torino 1982

Crovato, G. & M., *Regate e Regatanti.* Venezia 1982

Crump, T., 'The context of European Anthropology: the lesson from Italy' in J. Boissevain and J. Friedl eds, *Beyond the Community. Social Process in Europe.* The Hague 1975

Cucchetti, C.A. Padovan, A. and Seno, S., *La Storia Documentata del Litorale Nord.*. Venezia 1976

D'Ancona, A., ed., *La Leggenda di Vergogna. Testi del Buon Secolo in Prosa e in Verso.* Bologna 1869

————, *Le leggende di Vergogna e di Giuda*, in *Saggi di Letteratura Popolare.* Livorno 1913

Dante Alighieri, *Il Convivio*, ed M. Simonelli Bologna 1966

D'Onofrio, S., 'La Vergine e lo Sposo Legato.' *Quaderni Storici* 75, 859–878. 1990

Da Sera, D., *Opera Nova*, 1546 (1879)

Davanzo Poli, D. ed., *La Scuola dei Merletti di Burano.* Venezia 1981.

Davis, J., *People of the Mediterranean: an Essay in Comparative Social Anthropology.* London 1977

————,'The Sexual division of Religious Labour in the Mediterranean' in E.R. Wolf, ed. *Religion, Power and Protest in Local Communities.* Berlin 1984

Davis, M., *The Earlier Italian Schools.* London 1961

De Biasi, M., et al. *Storia di Burano.* Burano 1994

De Biasi M., ed. G.Nardo, *Studi sul dialetto di Burano.* Burano 1998

De Carlo, G., 'Beyond post-modernism', in Hatch, R., ed. , *The Scope of Social Architecture.* Wokingham 1984

De Felice, E., *I Cognomi Italiani.* Bologna 1980

————, *Nomi e Cultura.* Venezia 1987

De Lorris, G. and de Meun, J. (1237–) *Le Roman de la Rose* , D. Poirion ed.. Paris 1974, trans. Harry W. Robbins. New York 1962

De Mauro, T., *La Lingua Italiana e I Dialetti. Le Tre Venezie.* Firenze 1969–1972

De Martino, E., *Sud e Magia*, (1959) Milano 1972

Della Giovanna, E., *Il Tempo*, March 30 1967

Derelitti, Capitoli et Ordini. Venezia 1677–1678

Devoto, G., *Dizionario Etimologico.* Firenze 1962

Diena, M., 'Relazione della Deputazione Provinciale di Venezia presentata al Consiglio Provinciale. Atti'. AMV, *Comuni dell'Estuario:Unione a Venezia.* 1880

di Nola, A., *L'Arco di Rovo.* Torino 1983

Dionisotti, C. and Grayson, C., *Early Italian Texts.* Oxford 1949

Dorigo, W., *Una Legge Contro Venezia*. Roma 1973

Douglas, M., *How Institutions Think*. London 1987

Dresch, P., 'Wilderness of mirrors: truth and vulnerability in Middle Eastern Fieldwork', *Anthropologists in a Wider World* . Oxford 2000

Du Boulay, J., *Portrait of a Greek Mountain Village*. Oxford 1974

Dumont, L., *Essais sur l'Individualism*. Paris 1983

Ernaut, A. and Meillet, A., *Dictionaire étymologique de la langue latine*. Paris 1959

Earnshaw, P., *A Dictionary of Lace*. Princes Risborough 1982

Dunn, C.W., ed., *The Romance of the Rose*. New York 1962

Eliade, M., *Le Dieu Lieur e Le Symbolisme de Noeuds*. Paris 1952

Ellero, G., *L'Archivio IRE. Inventari dei Fondi Antichi degli Ospedali e Luoghi Pii di Venezia*. IRE. Venezia 1987

———, 'Personaggi e Momenti di Vita,' in *Nel Regno dei Poveri. Arte e Storia dei Grandi Ospedali Veneziani in Età Moderna. 1474–1797* , in B. Aikema, and D. Meijers. eds.. Venezia 1989

Enciclopedia Cattolica. Firenze 1954

Ernaut, A. and Meillet, A., *Dictionaire étymologique de la langue latine*. Paris 1932

Errera, A., *Atlante Statistico Industriale e Commerciale per il Veneto. Milano* 1871

Evans-Pritchard, E.E., *Kinship and Marriage Among the Nuer*. Oxford 1951

Fabian, J., *Time and the Other. How Anthropology Makes its Object*. New York 1983

Facco Delagarda, U., *Morte dell'Impiraperle*. Venezia 1967

Fambri, P., 'La Contessa Andriana Marcello' 1893. (Typescript)

Ferro, M., *Dizionario del Diritto Comune et Veneto*. 5 vols. Venice 1788–81

Filcea-CGIL, *Un Popolo di 'Verieri'. Mille Anni di Lavoro e di Lotte dei Vetrai di Murano e Venezia*. Venezia 1982

Filiasi, J.,*Memorie Storiche de' Veneti Primi e Secondi* , 11 vols., Venezia 1796

Fiume, G. and Scaraffia, L. (eds.), *Verginità,* Quaderni Storici 75, 701–14, 1990

Folena, G.F., *Gli Antichi Nomi di Persona*. Atti I.V. Sc. L. ed Arti. 1971 [cxxix]: 445– 484

Forgacs, D. and Lumley, R., eds, *Italian Cultural Studies*. Oxford 1996

Foster, George, 'What is Folk Culture?' *American Anthropologist*. Vol. LV, N 2. Part I, April–June 1953

Franco, G., *Habiti delle Donne*. Venezia 1614

Frankenberg, R., 'British Community Studies. Problems of Synthesis' in *The Social Anthropology of Complex Societies*. London 1966

Franklin, S. and Mc Kinnon, S., 'New Directions in Kinship Studies: a Core Concept Revisited'. *Current Anthropology* vol 41, Number 2, April 2000. p. 275–279

Franzina, E., 'Tra Otto e Novecento,' in S. Lanaro, ed., *Il Veneto*. Torino 1984

———, ed., *Storia Delle Città Italiane. Venezia*. Bari 1986

Galliccioli, G.B., *Storie e Memorie Venete Profane ed Ecclesiastiche*. Libro I, Capo v. Venezia 1795

Gambier, M., *Testimonianze sulla lavorazione del Merletto nella Repubblica di Venezia*. Comune di Venezia 1981

Gianichian, G.& Pavanini, P., *Dietro I Palazzi. Tre Secoli di Architettura Minore a Venezia. 1492–1803*. Venezia Arsenale 1984. 63–6

Gilmore, D. 'Introduction: The Shame of dishonor' and 'Honor, honesty, shame: Male status in contemporary Andalusia', in Gilmore, ed., *Honour and shame ad the unity of the Mediterranean*. Washington 1987

Ginsborg, P., *Daniele Manin and the Venetian Revolution of 1848–49*. Cambridge 1979
———, *Storia D'Italia dal Dopoguerra a Oggi. Societá e Politica 1943–1988*. Torino 1989
Ginzburg, C., *Il Formaggio e i Vermi*. Torino 1976
Giordano, D., *Relazione del Commissario D. Giordano sulla Amministrazione Straordinaria del Comune di Venezia*. Venezia 1924
Goddard, V. A. 1996, *Gender, Family and Work in Naples*. Oxford 1996
Goody, J., *The Development of the Family and Marriage in Europe*. Cambridge 1983
———, *The Oriental, the Ancient and the Primitive*. Cambridge 1990
Guiton, S., *A World by Itself*. London 1977
Hannerz, U., *Exploring the City. Inquiries Toward an Urban Anthropology*. New York 1980
Harré, R. and Lamb, R., eds., *Encyclopedic Dictionary of Psychology* : 'Guilt and Shame'. Oxford 1983
Herzfeld, M., 'Honour and Shame: Problems in the Comparative Analysis of Moral Systems'. *Man* 15, 339–51. 1980
———, *Anthropology through the Looking-glass. Critical Ethnogtaphy in the Margins of Europe*. Cambridge 1987
———, *A Place in History. Social and Monumental Time in a Cretan Town*. Princeton 1991
———, *Cultural Intimacy. Social Poetics and the Nation-State*. London 1997
Hine, D., 'Federalism, Regionalism and the Unitary State: contemporary Regional Pressures in Historical Perspective', in Levy, C. ed., *Italian Regionalism. History, Identity and Politics*. Oxford 1996.
Hirschon, R. , 'Open Body/Closed Space: The Transformation of Female Sexuality', in Ardener, S., ed., *Defining Females: The Nature of Women in Society*. London 1978
———, *Women and Property – Women as Property*. London 1978
———, 'Under one Roof: Marriage, Dowry and Family Relations in Pireaus.' In Kenny, M. and Kertzer, D., eds., *Urban Life in Mediterranean Europe: Anthropological Perspectives*. Urbana 1983
———, *Heirs of the Greek Catastrophe. The Social Life of Asia Minor Refugees in Piraeus*. Oxford 1989
Hirschon, R. and Thakurdesai, S., 'Society, Culture and Spatial Organization'. *Ekistics* 52: 15–21. 1970
Hirschon, R. and Gold, R., 'Territoriality and the Home Environment in a Greek Urban Community', *Anthropological Quarterly* 55, 53–73. 1982
Hodgkin, T., ed., *The Letters of Cassiodorus*. London 1886
Holmes, D. R., *Cultural Disenchantments. Worker Peasantries in Northeast Italy*. Princeton 1989
Honour, H., *The Companion Guide to Venice*. London 1965
Indovina, F., 'Le Trasformazioni Territoriali nell'Area Nord-Occidentale della Laguna di Venezia' in *Documentazioni di Architettura e Urbanistica*, (ed.) N. Ventura. Venezia 1975
Ingold, T. *The Perception of the Environment*. London 2000
Inquisition., Santo Uffizio. Processi. ASV, Folder 82 'Maria Zemella', 1626
Istituto Veneto Gramsci, *Per una Cultura della Trasformazione nel Veneto. Quaderno 1*. Arsenale, Venezia 1984
Jackson, A., ed., *Anthropology at Home*. (ASA Monographs 25). London 1987
Jacquard, A. and Segalen, M., 'Isolement sociologique et isolement génétique.' *Population* 3. 1973
Jamous, R., *Honneur et Baraka*. Cambridge 1981
Jesurum, M., *Cenni Storici e Statistici sull'Industria dei Merletti*. Venezia 1873

Just, R., *A Greek Island Cosmos.* Oxford 2000

Kertzer, D., *Comrades and Christians. Religion and Political Struggle in Communist Italy.* Cambridge 1980

——, *Family Life in Central Italy 1880–1910.* Rutgers 1984

——, *Ritual, Politics and Power.* Yale 1998

Kuntz, M. L., ed., *Postello, Venezia e il suo mondo.* Firenze 1988

Kuper, A., *The Invention of Primitive Society. Transformations of an Illusion.* London 1988.

Lanaro, S., ed., *Il Veneto,* in *Storia d'Italia. Le Regioni dall'Unità a Oggi.* Torino 1984

Lane, F., *Venice, A Maritime Republic.* Baltimore and London 1973

Lanfranchi, L. and Zille, G.G., 'Il Territorio del Ducato Veneziano dall'VIII al XII secolo, in *Storia di Venezia,* II, ch 1, 1–65. Venezia 1958

Lever, A., *Honour as a Red Herring.* Critique of Anthropology, Vol 6, N.3. 1986

Lepschy, L., Lepschy, G. and Voghera, M., 'Linguistic Variety in Italy', in Levy, C. ed., *Italian Regionalism. History, Identity and Politics.* Oxford 1996

Levi, E. *Il Libro dei Cinquanta Miracoli della Vergine.* Bologna 1917

Levi, U., *I Monumenti del Dialetto di Lio Mazor.* Arnaldo Forni Editore

Lienhardt, G., *Divinity and Experience. The Religion of the Dinka.* Oxford 1961

Lindisfarne-Tapper, N. and Ingham, B., *Language of Dress in the Middle East.* London 1997

Llobera, J. R., 'Fieldwork in Southwestern Europe: Anthropological panacea or epistemological straightjacket?' *Critique of Anthropology* 6 (2): 25–33, 1986

Loizos, P. 'User-friedly Ethnography?', in ed. Pina-Cabral, de, J., *Europe Observed.* London 1982

Loizos, P and Papataxiarchis, E., eds, *Contested Identities: Gender and Kinship in Modern Greece.* Princeton 1981

Lutyens, M., ed., *Effie in Venice. Unpublished Letters of Mrs. Ruskin. written from Venice between 1849 and1852. First edited by Euphemia Chalmers Grey, Lady Millais.* London 1965

Luzarche, V., ed.,*La Vie du Pape Grégoire le Grand .* Tours1857

Macdonald, S., ed., *Inside European Identities. Ethnography in Western Europe.* Oxford 1993

Maher, V, and Sellan, G., *Cucire e vestire: riflessioni antropologiche.* Quaderni. Quadrimestrale di psicologia e antropologia culturale. N, 3. Milano 2002

Malinowsky, B., *The Sexual Life of Savages in North-Western Melanasia.* London 1968

Manoukian, A., ed., *Famiglia e Matrimonio nel Capitalismo Europeo.* Bologna, 1974

Marchiori, C., ed., *Il Fiore e Il Detto d'Amore.* Genova 1983

Marcolini, G., 'Un Epilogo Settecentesco: Le Penitenti di San Job', in B. Aikema and Meijers, D.,eds, *Nel Regno dei Poveri.* Venezia 1989

Marett, R.R., *The Beginnings of Morals and Culture. An Introduction to Social Anthropology* (1931: 408).

Marsciani, F. 'Uno sguardo semiotico sulla vergogna', in Pezzini, I, ed., *Semiotica delle passioni.* Bologna 1991

Mozley, J.H., *The Art of Love and Other Poems.* London 1939

Marinetti, U., *Manifesti Futuristi.* (1910). London 1912

Marx, K., *Das Kapital* (1867), Chicago 1952

Marzemin, G., *Le Più Antiche Arti a Venezia.* Bologna 1938.

Mazzariol, G. e Pignatti, T., *La pianta Prospettica di Venezia del 1,500 disegnata da Jacopo de Barbari.* Neri Pozza. Vicenza 1963

Medin, A., *La Storia della Repubblica di Venezia Nella Poesia.* Milano 1904

Meldini, P., *Sposa e Madre Ideale. Ideologia e Politica della Donna e della Famiglia durante il Fascismo.* Rimini-Firenze 1974

Menez, F., 'La disparition des algues dans la lagune de Venise'. *La Ricerca Folklorica*, n. 42, 33–41. Ottobre 2001

Mewett, P., 'Exiles, Nicknames, Social Identities and the Production of Local Consciousness in a Lewis Crofting Community' in Cohen, ed., *Belonging. Identity and Social Organization in British Rural Cultures.* Manchester 1982

Miegge, ed., *Dizionario Biblico.* 1968

Miozzi, E., *Venezia nei Secoli.* Venezia 1957

————, La Verita sugli Sprofondamenti'. *Ateneo Veneto,* VIII, 109–120, 1970

Morris, C. *The Discovery of the Individual.* London 1972

Morris, J., *The World of Venice.* New York 1960

Mottola-Molfino, A., *I Pizzi: Moda e Simbolo.* Milano 1977

Muir, E., *Civic Ritual in Renaissance Venice.* Princeton 1981

Musatti, E., *Storia di Venezia.* Venezia (1914) 1962

Musolino, G., *Burano e il Suo Pastore.* Venezia 1962

Nardo Cibele, A., 'Studi sul Dialetto di Burano'. *Ateneo Veneto,* XXI, vol. 1, fasc. 3. Venezia 1838

Needham, R., ed., 'Introduction', in *Rethinking Kinship and Marriage.* London 1971

————, *Belief, Language and Experience.* Oxford 1972

Ninni. A.P., *Nozioni del Popolino Veneziano sulla Somatomazia.* Venezia 1891

Obici, Giulio, *Paese Sera,* March 19, 1967

Okely, J. and Callaway, H., eds., *Anthropology and Autobiography,* ASA Monographs 29. London 1992

Okely, J., 'Privileged, Schooled and Finished: Boarding Education for Girls,' in S. Ardener ed. *Defining Females* . London 1978

————, The Traveller Gypsies. Cambridge 1983

Ott, S., *The Circle of Moutains. A Basque shepherding Community.* Oxford 1981

Padoan, G., *La Commedia Rinascimentale Veneta (1433–1565).* Vicenza 1982

Pagan, Matteo, *Pianta Prospettica di Venezia, 1559* , ed., Francesco di Tommaso di Saló, Berlin 1567

Pagnozzato, R., 'Manichini veneziani 'da vestire', in V.Maher and G. Sellan, *Cucire e vestire: riflessioni antropologiche.* Quaderni. Quadrimestrale di psicologia e antropologia culturale. N, 3. Milano 2002

Pardo, I., *Managing Existence in Naples. Morality, action and structure.* Cambridge 1996

Parkin, D., ed., *Semantic Anthropology.* ASA Monograph 22. London 1982

Parkin, R., *Kinship. An Introduction to the Basic Concepts.* Oxford 1997

Parsons, A.'Is the Oedipus complex universal? A South Italian "nuclear complex", in R. Hunt, ed. Personalities and Culture. Austin 1967

Pasqualigo, G., *I Merletti ad Ago o a Punto in Aria di Burano. Richiamo Storico.* Trieste 1887

Pavanini, P., 'Abitazioni Popolari e Borghesi nella Venezia Cinquecentesca', In *Studi Veneziani,* N.S., Venezia 1981

Pellegrini, G.B. and Prosdocimi, *La Lingua Venetica.* Padova 1967

Pellegrini, G.B., *Saggi di Linguistica Italiana.* Torino 1975

Pemble, J., *Venice Rediscovered.* Oxford 1996

Peristiany, J. G., ed., *Honour and Shame: The Values of Mediterranean Society.* London 1965

Pina-Cabral, J. de, 'The Mediterranean as a Category of Regional Comparison: a Critical View', in *Current Anthropology,* Vol. 30, N. 3, June 1989, 399–406

————, ed., *Europe Observed*. London 1992

Pitt-Rivers, J., *The People of the Sierra*. London 1954

Parsons, A.'Is the Oedipus complex universal? A South Italian "nuclear complex", in R. Hunt, ed. Personalities and Culture. Austin 1967

Pocock, D., 'The Idea of a Personal Anthropology.' Decennial Conference of the Association of Social Anthropologists. Oxford 1973

Poppi, C., 'Building Difference: the political economy of tradition in the Ladin carnival of the Val di Fassa', in Boissevain, J., ed., *Revitalizing European Rituals*. London 1982

Postel, G., *Book of James, or Protovangelium. Venice*

Prati, A. *Vocabolario Etimologico Italiano*. Milano 1951

Pullan, B., *Rich and Poor in Renaissance Venice. The Social Institutions of a Catholic State,to 1620*. Oxford 1971

Putnam, R.D., *Making Democracy Work. Civic Traditions in Modern Italy*. Princeton 1993

Ravid, B., 'The Religious Economic and Social Background of the establishment of the Ghetto of Venice'. In, Cozzi, G., ed., *Gli Ebrei a Venezia*. Milano 1987

Reberschack, M., 'L'Economia', in Franzina, ed. *Venezia*. Bari 1986

Reiter, R. (ed.), *Toward an Anthropology of Women*. New York 1975

Rieu, E.V., trans. *The Iliad*, II. London (1950) 1977

Romanin, S. (1853) *Storia Documentata di Venezia*. Venezia 1972

Rompiasio, G., *Compilazione Metodica delle Leggi Appartenenti al Collegio e Magistrato delle Acque*. ed. G. Caniato. Venezia (1733) 1988

Rossi, F., *Patrocinio delle Donzelle Periclitanti. Statuto:* A I V. Venezia 1749

————, Will and Codicils, A I V. Venezia 1757, 1760, 1762

Rossi, V., *Le Lettere di Messer Andrea Calmo*. Torino 1888

Ruggiero, G., *The Boundaries of Eros. Sex, Crime and Sexuality in Renaissance Venice*. Oxford 1985

Ruskin, J., *The Stones of Venice*. London (1851–1853) 1858

————, *Praeterita*. London 1887

Salzano, E., 'Pianificazione Urbanistica'. *Comune di Venezia Bimestrale d'Informazione.* July–August 1979

Sanga, G., *Premana. Ricerca su una comunitá artigiana*. Silvana 1979

Saraceno, C. *Anatomia della Famiglia. Strutture Sociali e Forme Familiari*. Bari 1976

Savio, R. and L., 'L'organizzazione del lavoro femminile A Venezia nelle antiche istituzioni di ricovero e di educazione, in Mottola Molfino, A. and Binaghi Olivari, M.T., *I pizzi: moda e simbolo*. Milan 1977

Scano, L., *Venezia: Terra e Acqua*. Roma 1985

Schneider, J. and P. *Culture and Political Economy o Western Sicily.* New York 1976

Schneider, J. 'Of vigilance and virgins: honour, shame and access to resources in Mediterranean societies', *Ethnology* 10(1) 1–24

Scholem, G., *Jewish Mysticism*. New York 1961

Sciascia, L., *1912+1*. Torino 1986

Sciama, L.'The Problem of Privacy in Mediterranean Anthropology'. In Ardener, S., ed., *Women and Space: Ground Rules and Social Maps*. London and New York 1981

————, Kinship and Residence in a Venetian Island', *International Journal of Moral and Social Studies*. Vol 1, N. 2, 171–186. Summer 1986

————, 'Lace-making in Venetian Culture', in Eicher, J.B. and Barnes, R., eds., *Dress and Gender. Making and Meaning*. Oxford 1992

————, 'The Venice Regatta: From Ritual to Sport', in Mac Clancy, J., ed., *Sport, Identity and Ethnicity*. Oxford 1996

Segalen, M., *Historical Anthropology of thhe Family*. Cambridge 1986

Semi, F., *Gli 'Ospizi' di Venezia*. Venezia 1983

Serena, *Opera Nuova di Recami*. Venezia 1564

Signorelli, A., unpublished typescript. 1986

Silverman, S., "The Life Crisis as a Clue to Social Function' in Reiter R.R., *Toward an Anthropology of Women*. New York 1975

Spitzer, L., *Semantica Storica e Critica del Linguaggio*. Bari (1948) 1954

Sprigge, S., *The Lagoon of Venice, its Islands, Life and Communications*. London 1961

Stangherlin, A., *Storia Giuridica*, Venezia 1968

Stewart, C., *Demons and the Devil*. Princeton 1991

————, 'Honour and Shame', forthcoming in *Encyclopedia of Social and Behavioural Sciences*

Stewart, F.H., *Honor*. Chicago 1994

Strinati, R., *I Merletti ad Ago e la Scuola di Burano*. Roma 1926

Sutton D.E., 'Local names, foreign claims: family inheritance and national heritage on a Greek island'. *American Ethnologist* 24 (2), 415–37. 1997

————, *Memories Cast in Stone. The Relevance of the Past in Everyday Life*. Oxford 1998

Tagliavini, C., *Le Origini delle Lingue Neolatine*. Bologna 1949

Temanza, *Antica Pianta*. Venezia 1781

The Soncino Chumash, or Five Books of Moses. London, (1947) 1975

Tiepolo, M.F. et al., *Mostra Storica della Laguna* 1970

Toniolo, G., 'Sul Lavoro delle donne e dei fanciulli nelle industrie manufatturiere di Venezia e sopra alcuni criteri di legislazione industriale in Italia. Conclusioni del rapporto del comitato di studi economici di questa cittá'. in *Giornale degli Economisti*, Padova 1876

Tonkin, E., Mc Donald, M. & Chapman, M., eds., *History and Ethnicity*. ASA Monograph 27. London 1989

Trivellato, F., *Fondamenta Dei Vetrai. Lavoro, Technologia e Mercato a Venezia tra Sei e Settecento*. Roma 2000

Ughi, L., *Pianta Topografica di Venezia*. 1729

Ullman, W., *The Individual in Medieval Society*. Baltimore 1966

Unesco, *Rapporto su Venezia*. 1969

Vecchi, M., *Chiese e Monasteri Medioevali Scomparsi della Laguna Superiore*. 1983

Vecellio, C., *Degli Habiti Antichi e Moderni di Diverse Parti del Mondo. Libri Due*. Venetia (1598) 1590

Vianello, R., 'Il ruolo delle conoscenze etnozoologiche nella cultura materiale dei pescatori di Pellestrina'. *La Ricerca Folklorica*, N. 42, 27–32. Brescia: Ottobre 2001

Vico, G.B., *The New Science*, eds T.G. Bergin & M.H. Fish. Cornell 1968

Vocabolario degli Accademici della Crusca. Venezia 1612

Wikan, U., 'Shame and Honour: A Contestable Pair.' *Man* (N.S.) 19, 635–52. 1984

Williams, R., *Key Words*. London 1976

Wolf, E. R., ed., Introduction, *Religion, Power and Protest*. Berlin 1984

Wolfram, S., *In-laws and Outlaws. Kinship and marriage in England*. London 1987

Yanagisako, S. and Collier, J. F., 'Toward a Unified Analysis of Gender and Kinship'. *Gender and Kinship. Essays Toward a Unified Analysis*. Stanford 1987

Zingarelli, N. *Vocabolario della Lingua Italiana*. Bologna 1962

INDEX

Index

249